U0183971

# 光 明 城

LUMINOCITY

看 见 我 们 的 未 来

谨以此书献给与项秉仁同行的

当代中国建筑探路者

这是一次很有意义的回顾与梳理。本书收录了29个设计作品，大多是由项秉仁先生亲自勾画草图、研究空间形体、推敲细节而完成设计全过程。在整理和编辑跨越四十年的图片和项目资料的过程中，我们感动于他与时俱进、不断接受挑战的创作生命力。无论是在城市、建筑、景观领域，还是对于室内空间的设计，项秉仁先生一直坚持以城市视野、美学素养、专业技术和丰富经验作为设计的支撑，并力求使设计符合使用者的行为习惯和审美感知。

这也是一段中国当代建筑历史的真实回顾，从一个侧面展现了一代建筑师的成长。书中文本体现了过去五六十年建筑观念和潮流的变化与演进，从中我们能对当代建筑学所处的状态有更好的体会和认识。而在另一层面，不同于当代建筑界过多的表现主义欲望与个人英雄主义叙事情境，项秉仁先生的作品理性内敛兼具浓厚的人文情怀，能使读者感受到质朴的回归，促使人们不忘初心，回归建筑学本源进行思考。

——《项秉仁建筑实践1976-2018》编委会

Xiang Bingren Architectural Practice 1976—2018

# 项秉仁 建筑实践1976—2018

滕露莹 马庆禕 曹佟 李芸 编

上海·同济大学出版社 TONGJI UNIVERSITY PRESS

# 目录

# Contents

项秉仁

XIANG
BINGREN

建筑师
教授
建筑学博士
国家一级注册建筑师
美国加州注册建筑师
中国建筑学会会员
美国建筑师学会会员

Architect
Professor of Tongji University
Ph.D. of Architecture
Grade A Registered Architect of China
Registered Architect of California, USA
Member of the Architectural Society of China
Member of the American Institute of Architects

项秉仁 2016 年
摄于上海寓所

# 序言

# Preface

作为中国现代建筑的见证者，七十年来我陪伴了几代建筑师成长，他们之中不乏为城市面貌、学科发展作出卓越贡献者，也因此，我备感欣慰和自豪。其中，项秉仁是我一直很欣赏的学生。

落后于西方发达国家，我国始有真正意义上的建筑师是在20世纪20年代，第一代中国建筑师几乎都是留学外国学习建筑学的，而美国几个名牌大学的建筑系，尤其宾夕法尼亚大学堪称中国第一代建筑师的摇篮。项秉仁则是在南京工学院接受的建筑学教育，成为我国独立培养的第一位建筑学博士。“文化大革命”之后，我曾是他的硕士研究生指导教师，并继童寯先生接任他的博士生导师。项秉仁在本科阶段接受了完整的“布扎”学院派训练，基本功扎实，读书期间就受到老师和同学们的普遍关注。印象中那时他就非常关注当代世界建筑理论的发展和国际建筑大师的设计思想和实践，并出版了关于美国著名建筑大师的《赖特》一书，翻译了凯文·林奇的《城市的印象》。

项秉仁拿到博士学位后，他飞美国，去香港，再回到上海，在知名大学和设计机构学习经验，深入实践，逐渐成长为成熟的职业建筑师。今年，我收到他记录四十年成果的专著初稿，再次回忆起他的过人才华和端正态度，欣喜于他四十年如一日对于建筑设计的热爱、尊重，以及对创新的不懈追求。他是城市与建筑领域的优秀实践者，也是传播理想的教育者。他正视市场规律，也坚定地以更宏观、更有远见的专业身份去主导大型项目，保持了持续稳定的设计水准。从收录的二十多个项目中，可以看出他对于城与楼、人与景、古与今的诸多思考，他勇于打破常规，对于城市互融、可持续发展在实施层面做出了正面引导。

如今，我的学生项秉仁也已桃李满天下，我欣慰地看到一代代建筑学人的传承力量，中国现代建筑由学习西方为起点，必然要走向属于自己的东方道路。

齐康

中国科学院院士
东南大学教授、博士生导师
2019年8月

In the past seven decades of my life, I have witnessed the development of modern Chinese architecture, and had the pleasure to accompany the growth of several generations of architects. Many of these architects have made truly remarkable contributions to the appearance of cities in China as well as the progress of our discipline. For that, I am very pleased and proud. Among them, Xiang Bingren is one that I have always appreciated.

Different from his predecessor who had mostly studied architecture at renowned schools in the United States, especially the University of Pennsylvania and then started the profession of architects in China in the 1920s, Xiang received architectural education and was trained to be an architect in China in 1960s. After the "Cultural Revolution" Xiang return to Nanjing Institute of Technology for further study. I was his master tutor and doctoral supervisor after Tong Jun. Xiang was trained with Beaux-Arts system in his undergraduate years and paved his way with solid basic skills. Later in the postgraduate period, he made himself well known for being diligent and studious among teachers and students. As I recall, he was closely following the updates of contemporary architectural theories as well as design ideas and practices of international masters. He published a book about American modern architect Frank Lloyd, and also translated Kevin Lynch's *The Image of the City* during this time.

After receiving his doctorate, Xiang went on to expand his horizons in the United States, in Hong Kong, and then returned to Shanghai. Through learning and practicing at well-known universities and design institutions, Xiang gradually grew into a mature professional architect. When I received this monograph of documenting his forty years of achievements, remembering his brilliant talent and energetic spirit in his school years, I can't help but feel deeply his love, respect and unremitting pursuit of innovation. As a practitioner, he has faced up to the law of the market, and led large-scale projects with a macro and far-sighted professional affirmation, maintaining a stable design level. From the 20 plus projects included in this title, we can see his relentless contemplation on the city and the building, on people and society, on ancient and modern, and his courage to break the rules and provide positive guidance for the implementation of urban integration and sustainable development.

Now, Xiang Bingren, my student, has his students all over the world. What a privilege for me to actually see the inheritance from generation to generation. Modern Chinese architecture has started from learning from the west, but it's bound to move forwards along its own way.

Qi Kang

Academician of Chinese Academy of Sciences (CAS)
Professor and Doctoral Supervisor at Southeast University
August, 2019

# 起承转合

古人说字如其人、文如其人。优秀的建筑设计作品，应该也是建筑师本人的生活经历、专业立场和创作态度的体现。

对于项秉仁先生来说，就更是如此了。因为项先生在中国建筑师当中，有许多独有的经历，得到了很多项“第一”。他四十多年的创作过程，既是改革开放以来当代中国建筑教育和建筑实践的一个缩影，也是他独特的专业道路选择的结果。今天，项老师的作品结集出版，嘱我为序。余生也晚，作序当否，很是踌躇。然而项老师是我前辈师长，恭敬即是从命。于是不揣冒昧，借用“起、承、转、合”四个字作题一试。

## 起

项秉仁先生是中国内地第一个建筑学专业建筑设计方向的博士毕业生，在中国建筑教育界无人不知。他是1961级南京工学院（今东南大学）建筑学本科生，当年是南工建筑系有名的全优学生，作业在《建筑学报》发表过。他离校十余年辗转之后，1978年回到南工，师从刘光华、齐康、钟训正老师攻读硕士研究生。1981年硕士毕业后，又师从童寯、齐康老师攻读博士研究生，1985年博士研究生毕业。这样跨越十年后硕士、博士学位一气呵成的经历，在当时中国建筑界也是第一或者是唯一。随后项老师来到上海，在同济大学建筑系任教，成为第一批博士教师。后来又担任建筑设计教研室主任。这是一个起点非常高的教师生涯，足以让绝大多数的中青年教师羡慕。

还有一个起点，就是作品集里的第一号，马鞍山雨山湖公园小品建筑。这是在十年动荡的后期，在马鞍山设计院工作的项秉仁建筑师终于找到了设计实践的机会，1976年完成了对他来说非常有意义的第一波创作。这是在他攻读硕士、博士研究生之前的作品，已经能看出早期现代主义与中国园林建筑的结合。也许正是这个起点，推着项老师后来的一次次探索和尝试。

## 承

项先生的师承很了不起。杨廷宝先生教过他建筑初步，1966年项先生毕业时差点成为杨先生的研究生。有人认为，南京工学院的建筑教学体系是“布扎”式的学院派，重基础训练，重立面形式。也有人认为，在20世纪60年代社会背景下，在“实用、经济，在可能的情况下考虑美观”的原则下，南工的布扎和同济的包豪斯，应该都是打了折扣的。在硕士研究生阶段，刘光华、齐康、钟训正三位大家指导项秉仁等三位研究生，在刘老师的指导下项秉仁硕士阶段以赖特为研究对象，还翻译了凯文·林奇的《城市的印象》。而童寯先生，则要求项秉仁博士阶段把《古文观止》的某些篇章翻译成英文，这是独到的打通中外、古今的教学方法，也许这就为项秉仁先生以后的发展打下了基础。

除了南工的杨廷宝、童寯、刘光华、齐康、钟训正诸位见面的先生外，项老师没见过面的老师就是认真研究与翻译过作品的赖特和林奇。后者对项老师完成博士论文《城镇建筑学基础理论研究》，以及从事城市设计实践与研究有很大的影响。1987年的胡庆余堂药业旅游区规划，以及后来2006年的杭州元福巷历史街区保护更新，应该可以反映出这样的师承。

## 转

项秉仁先生来到同济，和老师们、同学们相处得非常融洽。南工的校友觉得他很南工，本、硕、博都是南工的，跨度长达24年；而同济的教师觉得他很同济，亲切随和，跟年轻教师打成一片。当然这跟他小学、中学在上海读书，能说一口地道的上海话也有关系。其实南工的优秀校友到同济任教历来是有传统的。在项秉仁老师之前，有戴复东、陈宗晖、卢济威、顾如珍等先生；在项秉仁老师之后，有常青、李浈、童明、李立、陈泳、胡滨、张永和等老师。南工和同济，是来往最多的。

同济的国际合作氛围一向是比较好的，贝聿铭先生数次来访并担任名誉教授，我记得当年项老师协助戴复东先生，跟两位德国建筑师合作做某个国际竞赛项目，有不少中青年教师参加。项老师1987年完成了具有实验意味的昆山鹿苑市场。

项老师离开同济的时候，我刚刚留校任教半年。那时学院只有一栋红楼，教师们不分老少，经常可以见面，也经常跟项老师打招呼，有说有笑的。忽然就听说项老师去美国了。后来才知道，项老师当时抱着“作为建筑师，45岁前一定要出国看看”的想法，走出了自己的舒适区，到美国作访问学者，并且在事务所兼职设计工作。人到中年，初到美国，不易之处可以想见。但项老师凭着自己的能力和勤奋坚持下来了，还通过了美国加州注册建筑师考试并获得执业资格；又获得了贝聿铭中国学者旅美奖学金。1992年，项老师转往香港，开始了新的职业建筑师生活。

这个转型有三层意思。空间上，从熟悉的上海转往美国，再转往香港，从语言到生活方式都有很大的变化；身份上，从大学教授变为开业建筑师；从创作态度上，从实验性建筑师转向了职业建筑师。1998年设计的江苏电信综合业务楼，就与十年之前的鹿苑市场全然不同——不仅是形式，而且是立场。在香港工作的七年，是忙碌的、复合的、多元的，“什么都要管”，远比当年的教研室主任更加复杂。这也许就是项老师当年隐隐之中向往的“建筑师的工作和生活”？我也听说，学校和学院领导多次邀请项老师回到学院来。

## 合

正好十年之后，1999年项老师回到了上海。他担任同济大学建筑与城市规划学院教授、博士生导师，并被聘为设计方法团队责任教授，随后他创立了上海秉仁建筑师事务所。在学院任教直到退休的十来年里，项老师培养了几十名硕士生、博士生，以及为本科生上设计课。他对学生既有很高的要求，又尊重学生自己的兴趣和选择，还一如既往地保持着友善和热情。

他对新兴的设计技术非常有热情和兴趣，例如数字设计等，今天在数字设计领域颇有声誉的袁烽教授出自他担任责任教授的团队。项老师在这一阶段更加明确了自己的设计理念：新现代主义（Neo-modernism）。比较有代表性的作品是2003—2009年的合肥大剧院，表达出“理性中的浪漫”。还有2008—2012年的宁波文化广场，不仅在城市设计方面探索了规律中的丰富和多元，而且在建筑的形式语言上有了新的发展。

项老师多位过去的学生成了他现在的合伙人，项老师也从事务所日常的事务中逐步抽身出来，让事务所里几十名年轻人有更多施展才能的空间。这也是这本作品集只截至2018年的原因：书中所录作品都是项老师亲自设计的。

到这里，建筑学教授和职业建筑师合一了，探索和日常合一了，现代主义和当代性合一了，老师和学生在一起合伙了。

我认识项老师三十五年了，他是一位热情平和、优雅大方的人，眼光敏锐，心胸豁达，兼有老一辈知识分子的讲究和年轻一代建筑师的轻松。我不记得他有喜怒形于色的时候，也从不觉得他比我年长二十岁。他对年轻的同事和学生特别宽厚，一直是我努力学习而遥不可及的榜样。

我最熟悉的他的作品是复兴公园南门改造。椭圆形的广场空间宜人，不露痕迹；栅栏和大门的比例舒服，细节讲究；金属、石材和少量的玻璃相互搭配，点到为止，这是对上海地域文化的理解和诠释，也是对他从小生活过的复兴路最好的回报。

项秉仁老师建筑实践的四十年，是起承转合精彩的四十年！

李振宇

同济大学建筑与城市规划学院院长、教授

2020年1月28日凌晨

# Qi, Cheng, Zhuan, He

## From the Beginning to the Present, The Track of Xiang Bingren's Profession

As the old saying goes, style is the man himself. Good architecture mirrors the architect's own life experience, professional standpoint and creative attitude. This is especially true for Mr. Xiang Bingren, for he has the most extraordinary experiences among Chinese architects and owns many firsts to his credit. His professional development in more than 40 years is not only the result of his professional choices, but also a microcosm of contemporary Chinese disciplinary education and architectural practice since the reform and opening up. On the occasion of publishing his collected works, Mr. Xiang has kindly invited me to write a few words. As a fellow professional that comes after him, I feel honored yet reluctant. Nevertheless, his wish is my command. So here I am, trying to present at least a glimpse of Mr. Xiang's decades of work by borrowing four characters from Chinese classical writing, which in this context can be roughly translated as Qi, Cheng, Zhuan, He —"the beginning", "the inheritance", "the transition", and "the combination".

### Qi—The Beginning

Mr. Xiang is well known in China's architectural educational circle as the first Ph.D. graduate in architecture from a Chinese university. He enrolled in Nanjing Institute of Technology (NIT) to study architecture in 1961. There he excelled and even published his student work in *Architectural Journal*. In 1978, ten years after leaving school, he returned to NIT to pursue master's degree and studied from masters including Liu Guanghua, Qi Kang and Zhong Xunzheng. That was immediately followed by a Ph.D. study under the tutorship of Tong Jun, Qi Kang until he received the degree in 1985. Such an endeavor of finishing both master's and doctor's degrees in one go after a ten-year departure from school was unheard of in China's architectural circle. Later, Xiang came to teach in the College of Architecture in Tongji University, becoming one of the first doctoral teachers. Before long he began to serve as the director of architectural design teaching and research section — a very high starting point for a teaching career, enough to make the vast majority of young and middle-aged teachers envy.

This period also marked another beginning for Xiang: the first work in this collection, a small architectural feature in Yushanhu Park in Ma'anshan. It was during the late years of "Cultural Revolution" that Xiang, then working at Ma'anshan Design Institute, finally got the opportunity to practice, and completed his first significant project in 1976. This pre- MA & Ph.D. work was already showing a mixture of early modernism and Chinese garden architecture. Perhaps it was this starting point that fueled Xiang's later explorations.

## Cheng—The Inheritance

When Mr. Xiang graduated from NIT with a Bachelor's degree in 1966, he almost became a graduate student of Mr. Yang Tingbao who had taught him fundamental architecture. (Some people would say that NIT at the time was teaching Beaux-Arts system that focused on basic training and facade form. Others might argue that under the social background of the 1960s, under the principles of "practical, economical, and beautiful when possible", both NIT's Beaux-Arts system and Tongji's Bauhaus system would have been somewhat compromised). In his graduate years, together with two other peers, Mr. Xiang was tutored by three masters: Liu Guanghua, Qi Kang, and Zhong Xunzheng. Under the supervision of Mr. Liu, Xiang studied Frank Lloyd Wright and translated Kevin Lynch's *The Image of the City.* Mr. Tong Jun, on the other hand, asked his student Xiang to translate certain chapters of *Gu Wen Guan Zhi* — classical Chinese literary works— into English. This by no means orthodox method helped connect what's local and what's foreign, what's ancient and what's modern, and we might as well say, laid the foundation for Xiang's future development.

Apart from the aforementioned figures, Xiang also learned from two authors whose books he had meticulously studied and translated: Wright and Lynch. The latter had a great deal of influence on Xiang's doctoral dissertation "Basic Theoretical Research on Urban Architecture" and his practice in urban design thereafter. We can identify such inheritance in Hu Qing Yu Tang Pharmaceutical Tourist Area (1987) and Conservation and Renovation for the Yuanfuxiang Historic Block, Hangzhou (2006).

## Zhuan—The Transition

Mr. Xiang got along very well with faculties and students when he came to teach in Tongji. Alumni from NIT saw him as a schoolmate since he spent years studying there; other teachers from Tongji felt he was easygoing and fit right in. This was partially due to the fact that he could speak fluent Shanghainese, a benefit from his elementary and middle school years in Shanghai. But maybe more importantly, it has been a long-time tradition for outstanding alumni of NIT (later renamed Southeast University) to teach in Tongji. Before Mr. Xiang, there were figures such as Dai Fudong, Chen Zonghui, Lu Jiwei, Gu Ruzhen among others; after him, there are Chang Qing, Li Zhen, Tong Ming, Li Li, Chen Yong, Hu Bin, Yung-Ho Chang, etc. Both institutes have benefited from their long history of close collaboration.

Tongji is renowned for its welcoming environment of international collaboration. I. M. Pei, for example, had visited multiple times and served as an honorary professor. I remember there was a time when several young and middle-aged teachers, including Mr. Xiang, were assisting Mr. Dai Fudong in cooperating with two German architects on an international competition project. Meanwhile, in 1987, Xiang completed his experimental Kunshan Luyuan Market project.

And then one day he just left all this and went to the US. I had just started teaching in Tongji for six months when this happened. At that time, the college of architecture had only one red building to use, so basically everybody saw each other quite often. I remember Xiang's departure felt all of a sudden for me. It was until later that I learned at the time he was thinking "as an architect, I must go abroad to see before 45 years old". That's what drove him to walk out of his comfort zone. He went to the States as a visiting scholar and began to do part-time design work in architectural firms. It can't be easy for any middle-aged person to start anew in a foreign country, yet Mr. Xiang stayed, survived and thrived — he passed the Architect Registration Examination in California and obtained the qualification to practice. And he won the Pei Travel Scholarship as well. In 1992, he decided to move to Hong Kong and started a new career as an architect.

This string of changes is nothing but striking. In terms of physical space, he moved from a familiar city to the US then again to HK, which inevitably accompanied huge changes from language to lifestyle; In terms of identity, he shook off the label of a university professor and became a practicing architect; In terms of creative attitude, he shifted from being experimental to being professional. The Jiangsu Telecom Multi-functional Building (1998) is completely different from the Luyuan Market project ten years earlier, both formally and conceptually. Seven years of practice in Hong Kong was busy, complex, diversified and demanding, far more complicated than being the director of a teaching and research office. Perhaps this was the "work and life of an architect" that Mr. Xiang had yearned for. On the other hand, Tongji never stopped calling for him.

## He—The Combination

So he returned, in 1999, exactly ten years after he left. He started teaching in Tongji as a professor and doctoral tutor in the College of Architecture and Urban Planning, and was appointed as Professor-in-Charge of Design Methodology Team. For more than a decade until his retirement day, Mr. Xiang taught design courses for undergraduates, trained dozens of masters and doctoral students. Friendly and enthusiastic as always, he held high requirements for students, but also respected their individual interests and choices.

Mr. Xiang is very interested in emerging design technologies, such as digital design. Professor Philip F. Yuan, who is renowned for his achievement in the digital design arena, was a member of Xiang's design team. Professionally speaking, Mr. Xiang consolidated his neo-modernist concepts at this stage, through works like Hefei Grand Theatre (2003—2009), which expresses "romance in reason", and Ningbo Cultural Plaza (2008—2012), which not only explores the richness and diversity in the laws of urban design, but also develops further in terms of formal language.

Xiang returned to Tongji in 1999, and later founded DDB Architects Shanghai. Now with several of his former students have become partners in his firm, Mr. Xiang gradually withdraws from daily affairs in the firm and gives the floor to the younger generation. That explains why this collection of works is only up to a certain point: every work in here is attributed to the man himself.

It's also at this point that we are witnessing an innovative combination of exploration and daily norms, of modernism and contemporaneity. Now, an architectural professor and a professional architect have reconciled, and the masters and apprentices have merged into one.

Thirty-five years I have known Mr. Xiang, he is always warm, sharp and open-minded. He has the refinement of an older generation of intellectuals, and still the ease of a younger generation of architects. I don't recall ever seeing him outwardly emotive, and I so frequently forget that he is in fact 20 years older. He is so generous with young colleagues and students that I will always hold it as exemplary.

I am familiar with his works, especially the renovation of the south gate of Fuxing Park. That project showcases Xiang's unique understanding and interpretation of Shanghai's regional culture. The oval square space is pleasant without any intervention trace; the proportion of fences and gates is comfortable and the details are exquisite; metal, stone and a small amount of glass are used in such a restrained manner. This is his way of giving back to the place where he spent his childhood. And whee, what a life he has lived since then!

Li Zhenyu

Dean Prof. of CAUP (College of Architecture and Urban Planning) of Tongji University

January 28th, 2020

# 项秉仁的
# 建筑人生

# Life as an Architect
# of Xiang Bingren

## 青少年

1944年1月项秉仁出生于浙江省杭州市，5岁时跟随父母移居到上海，居住在当时的法租界辣斐德路（Route Lafeyette，今复兴中路）附近。这个地区人口稠密，华洋杂处，呈现出较典型的上海半殖民地城市风貌。虽然居住的是上海传统的石库门里弄住宅，但门外的城市空间和街景，特别是满街的法国梧桐都洋溢着一丝异国情调。不远处还可见到具有装饰艺术风格(Art Deco)的辣斐大戏院、红砖尖顶的基督教堂，以及与石库门里弄迥然相异的欧式花园公寓和重庆公寓，另外还有一处以法式几何图案为园林特色的法国公园（今复兴公园），而这片区域现今就是上海著名的新天地商业区和中共一大会议的会址所在地。或许正是从那时起，这种特殊的居住环境和文化氛围潜移默化地对幼年的项秉仁的审美取向产生了一定的影响。

项秉仁的整个青少年时代都沉浸在这样的氛围里。他当时就读的小学萨坡塞小学（现上海市黄浦区卢湾一中心小学），原来是法租界里一所以高档校舍和优秀师资著称的学校，在全市颇有名气。那里的老师善于开发和诱导孩童潜在的天智和能力，因材施教。童年的项秉仁就是在老师的启发和引导下对绘画产生了别样的兴趣。在老师的鼓励下，他将稚幼的画稿投送到当时颇为知名的《儿童时代》杂志社所主办的少儿美术作品大赛，一举获得大奖，自此开始对绘画产生了持续的热情，也开启了城市环境美学意识。内心对美的追求居然成为他在小学毕业后选择中学的一个主要原因。当时正值上海淮海中学新校舍刚刚竣工落成之际，这座矗立在上海淮海中路上的美丽建筑让项秉仁深深着迷，痴迷地想要报考这所学校，全然不关心这所新办学校的师资质量和距家远近。之后，他如愿以偿进入了淮海中学，也成为一名天天要乘坐公交车上学的初中生，而城市的画卷在他的视野里也逐渐展开。

项秉仁的高中学习生涯是在上海向明中学（原震旦女中）度过的。这所著名的高中地处上海传统的文化高地，附近聚居着不少文学家、艺术家，学校里也有许多学生出身艺术世家，自然带来一股挥之不去的艺术气息。同时由于这所高中的前身是震旦女中，不少教师也是多才多艺，潜移默化地影响着一届届的少年学子。而对于项秉仁来讲，向明中学富有魅力的校园和那座由邬达克设计的充满动感的流线形教学楼更富有吸引力，他与这座大楼共处了三年的青春岁月，颇有惺惺相惜之感。在这里他不但作为学生会的宣传干事继续施展自己的绘画才能为学校的校刊作美工，而且还担当起文娱委员组织了全校的话剧团、口琴队排练和演出。项秉仁在大学期间加入学校管弦乐团演奏长笛正是受到高中班主任老师的影响。可以说，向明中学的优质教育和艺术熏陶为项秉仁选择走上技术和艺术结合的专业道路，并最终成长为一名建筑师奠定了起点。

## 大学

1961年9月项秉仁踏进了南京工学院（今东南大学），在中大院那座严谨并略有些沉重感的西式古典建筑大楼开始了他的建筑学专业教育。这个初创于1927年的建筑学系此前已经经历了三十余年的摸索，在杨廷宝、刘敦桢、童寯三位教授的带领下形成了一套成熟的建筑设计教学的理论和方法。由于这几位前辈的学术背景（特别是杨廷宝和童寯在宾夕法尼亚大学的建筑教育背景），当时建筑系的教育很自然地继承着布扎（Beaux-Arts）的传统，在五年本科教育中建筑构图和表现方法的训练占了很大的比重，美术课甚至延伸到第四学年。严谨扎实的建筑基础训练，大量的素描水彩技法学习，古典建筑构图和渲染训练，建筑设计课对草图、形式、细部刻画和图面表现的专注等，为项秉仁日后一生的建筑观念和实践留下了至深的影响。他痴迷地爱上了这种建筑专业的基础训练，并深信：在建筑设计全过程中几乎没有什么问题不能用得心应手的描画来表达、交流和解决的。

大学期间在绘图

这种对于建筑艺术略带片面性的认识和理解，伴之以20世纪五六十年代来自苏联建筑的影响，一直延续到本科四年级修习建筑历史与理论课之时。在有限地接触到西方现代主义建筑大师的理论和实践后，他方才领悟建筑设计除了适用、造型、立面构图、虚实对比、比例尺度等之外，还有更多在社会和技术层面的考虑，包括“形式追随功能”“结构主义”“空间流通”“少就是多”等令他耳目一新的现代主义建筑思想，特别是柯布西埃的萨伏伊别墅、格罗皮乌斯的包豪斯校舍、密斯的德国馆和赖特的流水别墅为他打开了全新的视野，使他意识到原来没有什么事物是颠扑不破、不可挑战的，曾经被奉为金科玉律的传统和古典主义的建筑设计教条终将被声势浩大的社会变革浪潮冲垮，让位于更为科学和实用的现代主义建筑理论。此刻和以后的项秉仁似乎寻觅到一把开辟未来建筑设计之路的利剑和一盏引领前程的明灯。

出自对建筑学科的兴趣和热爱，项秉仁在本科五年里刻苦学习，并保持了一至五年级各科成绩全优的优异纪录，从此成为南工建筑系师生们口耳相传的“神话”。与此同时，为了调节繁重的学习压力和提升艺术修养，他不但积极地投入学校各种板报和美术活动，还参与了学校管弦乐队的排练和演出活动，担当长笛演奏员。然而，学习岁月并非总是风平浪静的，令人困惑的事情也会发生，当时如火如荼的“设计革命化运动”和“四清运动”对于建筑设计课堂造成干扰，评价学生设计作业的价值观日益混乱。在这场思想革命的漩涡中，项秉仁原先得高分的电影院设计作业突然被刻意贬为不及格。同时，在1966年毕业那年，项秉仁报考了在当时只有少数尖子学生才敢于报名的杨廷宝教授的硕士研究生，但随后不久爆发的“文化大革命”使这一愿望搁浅，不能继续求学也不能就业的他和其他同学一样滞留在学校，这样的生活一直延续到1968年。

## 初涉建筑设计实践

项秉仁在1968年之后的整整十年与国家的命运绑在一起，历经了激烈的社会动荡和多种人生体验。由于“文化大革命”的影响，他虽学业优秀却未被录取为研究生，也没有留校任教，而是被分配到地处武汉所属电力部系统下的中南电力设计院。这是一个服务于电站建设的工业设计院，建筑专业只是配套的辅助工种，尤其是在当时的国情下，建筑师并没有多少发挥的空间，加之项秉仁工作报到之际正值“文化大革命”派别之争的高潮，所谓的“停产闹革命”使得人人无所事事，新来乍到的他只好拿起画笔，为造反派绘制大幅宣传画聊以保持自己的艺术感觉，而内心深处则期待着拨云见日的那一天。

数年以后，历经南北辗转的项秉仁终于在安徽省马鞍山市建筑设计院重回到绘图桌前，开始了建筑师的专业工作。在这个新兴的钢铁城市，他设计并建成了各类建筑设计作品，涵盖居住建筑、公共建筑和园林景观建筑，如雨山区菜市场、王家山住宅楼、马钢医院制剂车间、马

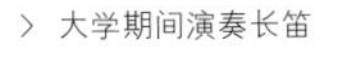
> 大学期间演奏长笛

< 大学期间在南京工学院建筑系馆前

> 大学期间在南京工学院大礼堂前

鞍山市电视台和马鞍山市建筑设计院大楼等。不难看出，这一时期项秉仁所做的建筑设计作品在观念上和方法上都是在追随和实践现代主义的建筑思想，如形式追随功能、空间流动、表现结构和简约立面等，而在技巧上仍体现出他在南京工学院建筑系所获得的来自布扎体系的基本功。1976年完成的马鞍山雨山湖公园建筑小品就是其中的一个实例。项秉仁基于当时当地的资金和技术条件，结合公园内水域景观的自然环境，运用现代主义的建筑设计手法和结构技术将建筑小品与其所在的基地环境融为一体，并且适度引入地方文化元素体现城市特色。在马鞍山市获得的建筑设计专业认可让他更坚信现代主义建筑理论的科学性和可行性，并认为似乎已经寻觅到了放之四海而皆准的金科玉律，然而此时的他身处于自我封闭十年之久的中国江南一隅，没有意识到整个世界的建筑发展早已如一江春水东流去，现代主义建筑“一统天下”的时代已然过去，一个令人眼花缭乱的、多元的建筑时代正在等待着他。

## 研究生

1978年国内恢复研究生制度，这使得项秉仁重新燃起了十多年前的愿望，经过努力他如愿回到了阔别已久的母校南京工学院建筑系攻读硕士研究生课程。当时学校十分重视“文化大革命”后的首批研究生，不仅坚持择优录取，而且在师资配备上也是投入最强的资源，刘光华、齐康和钟训正三位教授出任包括项秉仁在内的三位学生的集体导师。刘光华教授德高望重，是美国哥伦比亚大学建筑硕士，1947年回国，具有丰富的建筑教学和设计实践经验；齐康教授和钟训正教授也是当时建筑系的中流砥柱，后来分别成为中国科学院和中国工程院院士。入学以后老师结合国内外建筑设计的现状和学生们的不同背景经历布置了设计课的作业，并辅以精到的点评，学生因此获得了对于当时的建筑设计潮流的认识，建立起对建筑与环境、与城市、与历史文化的不可忽略的内在关系的理解，从而优化了自身的建筑价值观。除此以外的课程，以及中外建筑理论和实践的讲座、建筑实例考察、课外的专业阅读等，为曾经疏于专业理论学习的学生打开了新的视窗。

项秉仁就读硕士研究生的那几年正是西方建筑界全面反省现代主义建筑理论和实践的历史

时期，既有言之凿凿的分析批判，也出现了过犹不及、令人瞠目的后现代主义的设计案例。这一切无疑对曾经单纯信奉现代主义的他构成巨大的冲击，也引发他的反思。他既承认现代主义建筑理论固有的弊病，但又不能彻底放弃对现代主义建筑科学性的坚信。在这一时期的建筑作品中可以发现他明显的变化，同时也能看到些许的踌躇和妥协。项秉仁认为虽然后现代主义浪潮汹涌，但不免偏激和极端，这倒使他萌生了深入研究现代主义建筑大师的理论和作品的念头。项秉仁深感在本科学习期间，由于当时政治气候，学校教学对于现代主义建筑介绍十分欠缺又失之偏颇，他遂决定以美国现代建筑大师赖特的设计哲学和实践作为研究论文选题。通过大量原著的研读加深对现代主义建筑理论和实践的理解，从而也认识到现代主义建筑并不是千篇一律的教条，而是基于社会、经济和文化的不同呈现出不同的主张和表现，赖特正是其中的一个突出的例子。他最终在1981年完成了学位论文并通过了答辩。该论文几经充实修改后由中国建筑工业出版社在1992年作为“国外著名建筑师丛书”之一，以《赖特》为名出版，并在1997年被评为全国优秀出版物一等奖。对赖特的研究使得项秉仁加深和丰富了自己对建筑设计的理解从而逐步形成了他自己的建筑哲学，那就是建筑师应该通过认真的批判和分析来真诚地对待传统，应该继承和发扬自己民族的和地域的文化，尊重自然环境，尊重材料的本性，尊重时代，发展出一种整体和局部相统一的有机建筑。

完成硕士阶段的学习以后，项秉仁在选择就职单位时也遇到了一些挫折，茫然之际恰逢国家开始招收“文革”后首批博士研究生。当时建筑学科在全国范围内只批准了四位学者的博士导师资格，即清华大学的吴良镛、同济大学的冯纪忠、南京工学院的杨廷宝和童寯，而已取得硕士学位并愿意继续报考的考生也屈指可数。在老师与家人的鼓励和支持下，项秉仁最终成功地报考并被录取为南京工学院建筑系有史以来的首位博士研究生，导师是学术泰斗童寯教授。童老学贯中西，知识渊博，治学严谨，品格高尚，在建筑系师生中享有极高的声望。他对学生的中外文化知识和修养有很严格的要求，深感国内基础教育在古典文学和外语方面的欠缺，曾要求项秉仁熟读《古文观止》并能将其中的一些文章译成英文。他还让学生在图书馆阅读外文专业书籍并定期与他讨论，可以说童寯先生为项秉仁指引了从事学术研究和专业实践的正确道路。遗憾的是，中途

‹ 与导师齐康院士

› 博士论文答辩后与评审教授合影，1985年

童老因病仙逝，未能完成对项秉仁的博士阶段学习的全程指导。

齐康教授继任项秉仁的博士研究生导师，齐康先生建筑专业功底扎实，知识广博，尤其是具有建筑学与城市规划交叉学科的学术背景和修为，对国内外学术发展有深入的洞察和预知，在他的指导下，项秉仁开始以城市建筑为主题对于1960年代以来国际上关于城市设计、环境心理学、建筑符号学、社会学、人类学的研究进展进行了学习研究。与此同时他还在刘光华教授的帮助下翻译出版了凯文·林奇的《城市的印象》，这本在国际城市规划和建筑界享有盛名的重要理论著作，并且还撰写了多篇学术论文在《建筑学报》上发表。在这样的基础上，项秉仁最终选择了以“城镇建筑学基础理论研究”为题进行研究并撰写他的博士论文。在论文中项秉仁指出，在我国当前城镇建设中，建筑设计与城镇整体环境割裂的现象比较严重，突出的矛盾是城镇设计的理论研究与实践远远落后于现实的需要。鉴于此，论文通过对于国内外研究成果和实践的研究提出了城镇建筑学的理论框架，分析了城镇建筑环境的各要素和整体联系，论文强调提倡城镇建筑学会对我国的城镇建设和建筑教育带来有益的影响，在当前突出需要解决的问题是环境设计和文化理性。现在看来，三十多年前他的论文所提出的问题确实是有预见性的，对于日后项秉仁建筑观念的影响也是根本性的。

1985年10月，项秉仁完成了博士学位论文，并通过了由吴良镛教授为主席，由周干峙、郑孝燮、陈占祥、冯纪忠、戴复东、齐康、潘谷西等13位专家教授组成的答辩委员会长达5个小时的认真评审，成为了内地第一位建筑设计专业的博士生（在此之前清华大学的赵大壮通过了规划专业的博士论文的评审）。11月，项秉仁被同济大学建筑系聘用，一年后晋升为副教授。

## 旅美岁月

几乎整个1990年代，项秉仁是在境外度过的。1989年至1990年他在美国亚利桑那州立大学建筑与环境设计学院作访问学者，参与了设计教学和一个城市设计的课题研究，首次了解了在土地私有制的条件下进行城市设计的复杂性，体会到建筑师学习和掌握专业以外的社会知识的必要性。在此期间他有机会了解到，在亚利桑那州炎热干旱的沙漠地带，美国建筑师是如何结合当地的气候地理、民族文化和建筑传统去设计和创造符合现代美国人生活方式的当代建筑的。他访问了赖特的西塔里埃森设计营（Taliesin West），体会了这位美国建筑巨匠的独特魅力：包括赖特的建筑和自然共处共生的建筑哲学，赖特继承和弘扬美国本土建筑文化的强烈意志，他的建筑设计的无穷创造力，以及注重通过实际建造培养门徒的建筑教育思想和方法，等等。他也访问了建筑师保罗·索莱里（Paul Soleri）在菲尼克斯城北荒漠中兴建的城市实验室"阿科桑蒂"（Arcosanti），并为这位建筑师为了探求生态城市之路作出的近似宗教信徒般的执着努力而感动。在这段时期带着对于当地建筑师设计实践的强烈好奇心，他也临时性地参与了当地一些项目的设计，并学习到美国建筑师对于建筑专业的热爱和自豪，以及崇高的职业精神。

在结束了亚利桑那州立大学的短期访问之后，他先后在旧金山布朗·鲍特温事务所（Brown Baldwin Associates, San Francisco）和TEAM 7建筑师事务所参与当地各类项目的设计实践。1990年代中国内地没有建筑师注册制度，更没有私人执业的建筑师事务所，出于对在这方面已经具有成熟发展历史的美国建筑师事务所的浓厚兴趣，项秉仁决定加入其中亲身体验美国建筑师事务所的设计实践和运作方式。这段经历使曾长期囿于学术圈的他真正感受

< 在赖特设计的亚利桑那州立大学礼堂前，1989 年

> 在布朗·鲍特温建筑事务所，1990 年

到市场竞争的严酷和业主至上的绝对性——建筑设计是服务，建筑设计就是要解决问题，建筑设计必须要为客户带来利润等基本的专业伦理。而建筑师永远要以主动的态度，高层次和高水平地为客户找到解决问题的最佳答案。也许上述的困惑、体验、经验和信念混成了这一时期项秉仁的建筑思维并反映在其旧金山金融区高层写字楼的更新改造、太平洋贝尔公司总部、纳帕谷美国厨艺学院、烟台海滨规划和政府中心等建筑与室内设计作品中。

在美国的这段时期，项秉仁还获得贝聿铭先生专为中国访问学者提供的奖学金，这是贝先生把他获得的普利兹克奖奖金贡献出来作为基金，以资助来美学习的学者考察美国建筑，尤其是由贝先生设计的散布在美国各大城市的主要建筑项目。身历其中获得的体验和与贝聿铭先生的交谈给予项秉仁的是一次影响深远而且难以忘怀的建筑再教育。令他对于20世纪现代主义建筑的科学性、实践性、创造性和生命力有了更深刻的亲身体验和理解。在此后相当长的一段事务所实践中，他曾处于用现代主义的原则和方法解决项目的各种需求，用非现代主义的观点处理其余的设计问题的矛盾之中，他逐渐感到这两者其实并非那么势不两立，若善加处理则可以互相取长补短。现代主义建筑理论和方法能赋予项目设计理性，而非现代主义的思想能改善甚至消除现代主义的极端性和片面性。

与此同时，在准备和参与加州注册建筑师资格考试的过程中，项秉仁也逐渐深入了解了美国建筑规范和建筑项目管理制度的特点，并最终通过了加州注册建筑师考试的九个科目笔试和口试，取得了执业建筑师资格。

> 贝聿铭中国学者旅美奖学金，1992年

## 感知香港

1992年末，项秉仁听从贝聿铭先生的建议，离开美国投入到充满活力和动感的人口稠密的东方名城——香港，去品味一种东西融合的文化，体验一种以顽强意志克服天然和人为障碍而获得的生存环境。

香港这座城市和她的文化带给项秉仁最大的影响或许还在于使他认识到建筑空间的市场商业价值。建筑师和室内设计师所做的一切，似乎都是为了帮助和保证商人从最终消费者那里获得金钱回报。然而在这种压力下，建筑师的专业责任和职业尊严被尤为看重。他一直告诫自己永远不能只对一个客户，而是要为更广大的市民和属于他们的城市环境负责。这样的价值取向使得他在这一时期力求在建筑的商业价值和建筑品质追求之间寻取平衡。

在年近半百之际进入香港这种紧张和复杂的工作和人际环境实属不易，而这也促进了他全面建筑观的形成并使其个人建筑思考更趋成熟。他从混沌迷茫逐步走向豁然开朗——应该做什么建筑和能做什么样的建筑。

从北京国际金融中心规划、庄胜苑、湖南国际金融中心大厦、东方风情俱乐部、北京庄胜回迁住宅、上海龙柏苑、北京芳群公寓，直至后来的江苏电信业务综合楼、南京电视电话综合楼、南京泰山新村电信楼等项目的设计中不难发现这一时期项秉仁的建筑设计轨迹的发展。

香港人给他留下的另一个深刻的印象就是港人爱港。在历经数十年的奋斗后，香港已成为在软硬件环境建设方面绝不亚于世界任何地方的国际一流城市，而且又有着香港人同声同气的宜人的文化环境。这促使项秉仁在1999年决定返回上海。因为和许多香港人的想法一样，他的家、朋友和熟人在上海，上海乃至整个中国内地的发展使他感到发挥自己专长的机遇也在那里。

> 在香港主持 DDB 期间，1996 年

## 返回上海

1997香港回归前后，他亲眼目睹了一些海外归来的建筑师朋友在国内抢得先机，俨然成为业内的佼佼者，而他本人也获得不少的机会得以主持各种类型和规模的规划设计项目。在海外与外国建筑师同行的合作经历，最起码的收获是知己知彼，树立了自信心。他深深地意识到，抛开体制上的差异，内地的建筑师并不逊色。十几年的飞速发展使内地同行既开了眼界，也受到了磨炼，已经能胜任设计高水准的建筑项目。他想与他们一起努力来证明，这也成为他回国发展的强大动力。

1999年9月，项秉仁回到上海成立了项秉仁建筑设计咨询有限公司（DDB International, Ltd.），而后成为上海秉仁建筑师事务所（DDB Architects Shanghai）。刚开始他参与一些小型的城市更新项目，而这些项目也开启了他日后采用城市视角、遵循历史脉络、富有人文情怀来看待建筑设计的建筑观。同年他被同济大学建筑与城市规划学院聘为教授和博士生导师，任建筑设计方法学科组责任教授。作为知名建筑学科的众多教授中的一员，项秉仁凭借多年来在建筑设计领域和大学学术环境的经历，切身体会到一名合格的建筑设计与理论学科教师，首先要有一定的建筑设计工程实践经验，其次是要鼓励学生和同事们发现和发掘自己的学术兴趣和研究课题。因为他认为目前唯有大学才是学术自由的园地。他的学科组成员都在各自的研究方向和实践中，取得了傲人的成果。而他所指导的研究生，特别是博士研究生各自选择了感兴趣的论文选题，毕业后也都在擅长的专业领域成绩斐然。他也因此多次获得学校颁发的优秀教师奖。

经济发展带来了国内建筑设计界前所未有的活跃。这表现在许多方面，如建筑设计队伍的日益多元化，建筑设计学术研究水平的不断提高，中外建筑同行的交流和竞争增多，媒体和出版物的积极推介以及社会各界的关注，等等。在这种氛围中项秉仁既受到推动和激励，也感受到实实在在的压力，体会到在专业发展的道路上是不可能一劳永逸的。面对和解决任何新的问题不仅需要经验和积累，还需要不断的学习和借鉴。与此同时，在周围一片喧嚣声中也特别地需要冷静而清醒的独立思考：怎样才能真正建设好我们的城市？在弥补过去的缺憾、大力提倡城市规划和城市设计的同时，作为建筑师应如何进一步提升本身的理论修养和专业技巧，专注于把建筑设计做得更精更好，以能够真正实现城市设计的目标。在面对沸沸扬扬炒作的建筑的文化、历史、象征、新颖、时尚等建筑外在的因素的同时，是否可以以平实的心态去认真踏实地逐一解决所接受项目的环境、功能、流线、空间、材料、技术和造价等问题？在检讨迎合市场大众口味的同时，是否应倡导一种朴实而健康的建筑审美，深入挖掘建筑物固有构成要素的美学潜力，包括它的空间、形体、技术、材料、细部、色彩、光影、虚实等，从而为大众营造平静、优雅、尊重生态和自然的高品位的物质环境。自1999年回到上海以后，项秉仁逐渐步入眼花缭乱的建筑设计市场，得到了许多机会，项目涉及许多城市设计和建筑设计类型，他的事业迎来一个前所未有的黄金时代。

## 职业征途

自从项秉仁先生1961年踏入南京工学院建筑系的大门以来，迄今已逾半个世纪。在这段人生中，项秉仁在国内接受了良好的本硕博建筑学教育，经历了“文化大革命”，感受了不同地域文化背景的巨大差异，其余的时光都全身心投入到建筑教学和设计实践当中。他曾作为访问学者到访过香港大学、东京大学、亚利桑那州立大学、韩国密阳大学等建筑院系，也在美国旧金山和香港度过了共计十年的执业建筑师专业生涯。自从在1999年回国后，他凭借其国际化视野、良好的专业素质和多年的职业经历开创了自己的事业。2012年，项秉仁荣获中国建筑学会授予的“当代中国百名建筑师”荣誉称号。

项秉仁将人生各个阶段的经验，不同的学习与工作经历转化为建筑设计的基础修养。他在国门再次打开后纷至沓来的各种建筑学风格、思想与流派的表象后摸索普适的建筑学基本规律，在不同的文化与社会背景的流转中坚定建筑师的职业操守。项秉仁的建筑学专业求学之路，以南京工学院新古典主义建筑教育的布扎体系为基础，这种严谨苛刻的训练方法奠定了项秉仁初为建筑系学生时的审美取向和价值判断，而高年级讲授的现代主义建筑大师的建筑理论和方法则对他产生了长期的，甚至是终身的影响。这两者构成了项秉仁建筑创作观念和方法的基本面。但是必须强调的是，项秉仁对每个项目的设计出发点是各不相同的，其设计方法是在全面掌握项目相关资料后竭力思索，分析比较再优化结果，追求的是独特性和创意。为每个项目提出的问题寻找到特定答案是他的最大成功和乐趣，也是他的建筑设计之道。

项秉仁的建筑生涯横跨了改革开放前后，其建筑设计思想的成熟与中国建筑学重新融入全球建筑潮流是同步的。在改革开放之前的设计实践中，项秉仁已经开始协调布扎体系的形式原则与现代主义的科学方法之间的冲突。改革开放后，在诸多后现代主义建筑思潮不断冲击经典现代主义建筑范式的年代，项秉仁较早认识到现代主义的本质是对质量、标准与价值的追求，以及建筑学价值体系的普适性与特殊性在设计过程中各自适用的领域。在此基础上，项秉仁建立起处理现代主义基本设计方法与复杂的文化维度之间关系的独特方法。在之后的海外学习实践经历中，项秉仁开始建立现代建筑师的职业观，顺利完成了向一个现代意义的独立执业建筑师的转型。职业观（professionalism）是现代主义建筑思想不可缺少的成分与隐含前提，界定了建筑师在整个建筑设计过程中的社会角色。项秉仁以自己的实践验证并发展了现代主义建筑学内在的职业观，影响了一代年轻建筑师群体的职业道路。

评论 Review

谭峥
Tan Zheng

# 后学院派时代的“新现代主义”

## 项秉仁建筑学思想初探

# Neo-Modernism in Post-Academic Era: Exploring the Architectural Thoughts of Xiang Bingren

谭峥，建筑学博士，同济大学建筑与城市规划学院助理教授。

## 项秉仁与中国建筑的"当代史"

1961年,17岁的项秉仁进入南京工学院(现东南大学)学习建筑学。此时的中国建筑学教育正进入一个小阳春。得益于1960年代初比较宽松的政治环境,杨廷宝、童寯等第一代建筑学者能够进一步延续得自宾夕法尼亚大学,并从东北大学时期就开始贯彻并改造的"布扎"建筑学教育体系。这一体系以"图房"(atelier)为组织形式,以徒手草图能力(esquisse,或可翻译为"快图")的训练为核心,强调扎实的美术功底,致力于培养能够独立完成设计的职业建筑师。这一学院派体系的黄金时期一直延续到1966年。于是,以项秉仁为代表的一代建筑师成为新中国所培养的最后一批完整的接受"布扎"学院派训练的建筑师。十余年以后,即便建筑学教育重新步入正轨,"布扎"体系也不复往日旧观。此时,"当代"的大幕拉开,建筑学教育正式进入"后布扎"时代。

与"布扎"教育的淡出、中国的当代建筑学的显形相始终,项秉仁的建筑学探索之路可以粗略分为三个阶段。第一阶段从在南京工学院的建筑学教育开始,到1985年获得博士学位为止,当中经历了"文革"的波折,是项秉仁建筑观的孕育与激荡的时期;第二阶段从第一次在同济大学任教开始,到接受贝聿铭旅美奖学金为止,涵盖了项秉仁从青年先锋向成熟建筑师的转变,是他的建筑观的自我赋形时期;第三阶段从赴香港继续建筑学实践开始,到回到上海后的十余年实践为止,是他的建筑观逐渐稳定并探寻新的自主性的时期。本文之后的写作将以此为线索,次第展开。

西方建筑学的当代史界定问题基本上音调已定,即将1960年代的新先锋主义运动视为西方现代建筑史的终结,而将1970年代开始的后现代主义思潮(及随后发生的解构主义、生态主义、新现代主义、参数主义、批判的地域主义等)视为当代建筑史的发端。与之相似的是,中国建筑的"当代史"在近年的研究中日益成为一个问题。"当代史"的问题可以表述为:哪些历史进程与事件构成了目前建筑学的样貌?以社会史的视角,改革开放之后的建筑史可以被视为当然的当代史。但是,我们应当看到,"当代"并不是随着社会变革而自然发生的,因为建筑学的教育要作用于实践,需要一个孕育、孵化与蝶变的过程。对当代建筑史的研究不能囿于意识形态领域的变迁,建筑学本身的生产与运作机制是理解当代建筑发生的关键。

史建认为,新千年后的"实验建筑"从个人的、观念的空间叙事,逐渐全面蝶变为更广泛的"当代建筑"。但是在"实验建筑"向"主流建筑"的转化之外,建筑学在更广阔的社会与生产机制中发生了怎样的转型?如何在观念史以外解释"当代建筑"的发生?在改革开放后的二十年中,对于一种熔铸身份、场所与历史的中国式现代建筑的追求固然是主流之一,然而对于更普适、优雅的、职业化的建筑学语言的探索也悄然进行着。这种设计方法客观上柔化中和了不同的形式元素所携带的观念冲突,超越历史、立场与身份的争辩,构成了新千年的中国大都会建筑形象的基色。本文试图将这种建筑设计思想定义为中国式的"新现代主义"(Neo-modernism ),在中国的语境下,"新现代主义"是各种当代建筑学思潮与风格激荡冲突中较为精致、收敛且调和的那一部分,它与西方晚期现代主义建筑所呈现的面貌有相似之处,但在中国更倾向于抑制激烈的形式元素冲突,更能代表当今都会的职业中坚力量的价值取向,最终成为更具广泛代表性的当代建筑的一部分。

在"当代史"的界定与研究中,有一些问题依然悬而未决,对项秉仁先生的建筑思想的研究与这些问题关系密切。首先是重新认识"布扎"传统在当代中国(后改革开放时期)的改造、延续,发展与变异。如何在中国现当代建筑历史与批评的角度对"现代主义"与"布扎"的关系进行再认识?它们究竟是一种思潮,一种风格,一种方法还是一种修养?如果有一个"后布扎"时代,那么它的特征是什么,它从"布扎"继承了什么?学科史意义上的更强调建筑学自主性的伦理与职业环境中更强调社会反馈的伦理究竟有何异同?

其次,如何界定项秉仁先生这样最后一群完整地接受"学院派"训练的建筑师在中国当代建筑史中的作用?项秉仁建筑师作为当代建筑师的范例的典型性与特殊性在哪里?他相对于新浪潮或"实验建筑"的态度是怎样的?他后来逐渐远离青年先锋色彩的抉择是否在主动地保持距离,还是在各种即时即刻情势下做出的选择?

此外,项秉仁先生另一重要建筑学设计思想的转型大体完成于香港,这一阶段他直接受到香港本土的"新现代主义"潮流影响。在他的眼里,香港具有一种超越地域性的文化特征。那么,中国内地、香港与美国各自代表了怎样的建筑职业文化?这种文化地理上的维度对书写中国当代建筑史有何助益?从香港等后发亚太地区都市迁徙来的"新现代主义"是否代表了中国当代建筑文化的某种价值取向?从香港回到上海以后,在最新的实践中,在探索数字化时代建筑学的状态的过程中,项秉仁的建筑设计进入一种新的自由境界,那么哪些作品可以集大成地代表这种自主性?这种自主性是否是他之前四十年的建筑学思想的延续与发展?

## 分代

"分代"是中国当代建筑研究中既武断又高效的分析方法。根据杨永生先生的《中国四代建筑师》,现代意义上中国建筑师可粗略分为四代:第一代出生于清末到辛亥革命间,以受益于庚款留学制度的归国留学英美毕业生为主(梁思成、陈植、童寯、陆谦受等);第二代出生于20世纪10年代到20年代,在1949年前学成执业(冯纪忠、王大闳、黄作燊、刘光华等);第三代出生于20世纪30年代到40年代,在"文革"之前接受建筑学教育(何镜堂、程泰宁、布正伟、齐康等),在改革开放后成为塑造当代中国城市面貌的主要力量;第四代出生于1949年之后,在恢复高考之后接受大学教育(张永和、刘家琨、王澍等)并在20世纪90年代迈入历史舞台。

这一年代标准也与重大的历史变革相对应。在这一分代法的基础上，彭怒和伍江在《中国建筑师的分代问题再议》一文中将第三、第四代建筑师受教育的分界点定在1978年，即将恢复高考后接受建筑学教育的建筑师定为第四代，这基本上确立了四代建筑师的划分原则。

这一中国建筑师分代原则提供了若干分析当代建筑师的方法，如教育、师承、社会背景等，所以它的意义并不在分代，而在理解个人与历史大背景之间的关系。如果兼顾具体的个人阅历经验，则项秉仁先生同时具有第三代与第四代建筑师的特征。1960年代，新中国培养的第一批建筑学专业人士（即第三代建筑师）已经走上教育岗位，这包括齐康、钟训正、孙钟阳等，项秉仁先生与第三代建筑师中的代表——齐康先生，有着亦师亦友的关系，研究生阶段又与第四代建筑师的中坚——“新三届”（孟建民、张永和等）的学习经历发生重叠[1]。在1980年代，随着第二代建筑师退出历史舞台，建筑学的中流砥柱已经是第三代建筑师。随后，在美国与香港进行建筑设计实践的过程中，项秉仁与一些海外华人建筑师群体有着生活与工作上的交集。这一系列复杂的经历淡化了项秉仁的代别特征。

分代原则对于理解个人与群体共性相关的特征显得便捷有效，而对理解个体在具体情境中的行为与抉择，却少有助益。整个1980年代正值世界主流建筑思想全面向国内输出的时期，当时的建筑师经历了多次建筑思潮和建筑审美的变革，这一思想的激荡在项秉仁的成长过程中尤为突显。项秉仁在特定历史情境下作出了数次具有重大影响的抉择——从最初报考南京工学院建筑学专业，到选择攻读博士，再到赴美国深造学习，最后回到国内继续建筑设计实践。这些抉择不仅是历史情境下的反应，而且是自身的禀赋与个性在某些偶然机遇下的表达。

## 第一阶段

### “学院派”与现代主义建筑

自1961年到1985年，除了十年“文革”的中断，项秉仁先生在南京工学院接受了12年的建筑学教育。1961年到1966年之间，政治运动对建筑学教育的干扰较少，这令项秉仁能够在本科阶段完整地接受了“布扎”体系的“学院派”教育，且在本科学习的后期接触到现代主义建筑思想。“文革”后，作为首批建筑学硕士及博士研究生，项秉仁又迅速接触到涌入国内的各种西方建筑学思潮。项秉仁的成长史与建筑学教育的布扎体系、包豪斯体系、后现代主义与城市建筑学等诸思想流派在“文革”前后二十年间的碰撞、演变与消长是同步的。这些教育方面的影响依次发生，分别界定了项秉仁的基本建筑学修养（布扎体系），深层的建筑学信仰（现代主义）与包容的、适应性的建筑学视野（后现代主义与实用主义）。

1956年全国高等院校院系调整，建筑系随南京大学工学院在原中央大学旧址成立南京工学院建筑系，伴随着南京工学院毕业生向其他地区建筑院校的输出，“学院派”建筑教育在全

> 本科期间课程作业被选送至《建筑学报》发表，1964 年

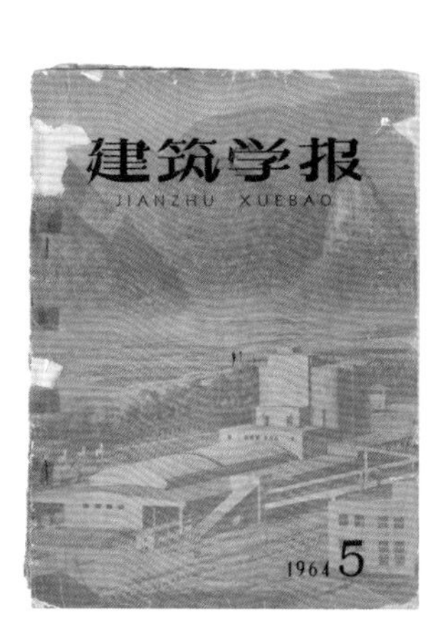

国影响日趋强盛。南京工学院也成为当时有志于学习建筑学的青年学子的“圣地”。此时的中国建筑学教育开启了1949年后的第二个黄金时期。1962年，陈毅在广州会议上发表了为知识分子“脱帽加冕”的讲话，检讨了前一段时间中央对知识分子的态度，梅兰芳式的“又红又专”的艺术家成为知识分子的楷模，兼之国民经济处于调整期，“左”的影响对科教文艺的干扰开始淡化。在南京工学院，专业教育的核心地位被重新确认，教授学者的权威也得以树立。

项秉仁的艺术启蒙在中学阶段就已经完成，在上海向明中学时，他就已经是文艺干事，组织了全校的话剧团、口琴队排练和演出等，后来在大学期间又参加学校管弦乐队演奏长笛。项秉仁一直想学习一门结合艺术与技术的专业，在报考南京工学院之前，他曾经有赴北京电影学院学习摄影的打算，但是未获录取，因此最终选择了南京工学院的建筑学专业。项秉仁先生的文艺才华对其早期的建筑学专业学习起了极大的辅助作用。当时的建筑学教育极其重视绘图的基本功训练，包括素描、水彩、渲染、测绘、阴影透视等，直到四年级还在画水彩画，美术基础扎实的项秉仁迅速成为专业能力突出的明星学生。据项秉仁先生回忆，除了强调绘图训练外，当时的核心课程也包括中外建筑史、城市规划、建筑施工、建筑力学、建筑结构、建筑技术、给排水和电气工程等。项秉仁的各科成绩均为优秀。

这一时期他留下了众多的设计习作，多数习作以正统的学院派黑白水彩渲染表达，习作内容包括火车站、学校等，许多习作已经存档并成为后来的学生们学习的范本。在四年级时，项秉仁在刘先觉先生的外国建筑史课上第一次了解到现代主义建筑的概况，这包括“四大师”的理论与包豪斯的主张。项秉仁对当时“中而古”的正统并不热衷，却对现代主义建筑产生了浓厚的兴趣[2]。恰巧在1964年，清华大学的吴焕加发表了《评西方十座建筑》一文，介绍并“批判”了《建筑论坛》杂志评选的“六十年代的十座伟大建筑”(其中7座在美国，意大利、巴西、法国各1座)。这些形式各异的晚期现代主义建筑对当时处于学院派主导的国内建筑教育产生了一定的冲击。这些文章虽然以“批判”的面目出现，事实上宣传了现代主义的理念。自此以后，项秉仁开始主动接受现代主义建筑，也在设计作业中遵从以功能主导的现代主义建筑学设计方法。项秉仁了解到同济大学的黄作燊与冯纪忠教授的现代主义教学与设计理念，但是一直没有机会看到实物，后来终于在“文革”期间得以到同济实地参观。

“文革”开始后，项秉仁受到冲击，被认为是资产阶级教育路线培养的“反面典型”，既不能读研也不能留校，无奈之下辗转湖北武汉、辽宁鞍山与安徽马鞍山的不同工矿企业，最终在1973年进入马鞍山建筑设计院。这个新兴的钢铁城市有着大量的建设任务，项秉仁也得以负责

< 出版译作《城市的印象》，1990 年

> 出版著作“国外著名建筑师丛书”之一《赖特》，1992 年

荣譽證書

项秉仁：

荣获 当代中国百名建筑师

特颁此证

> 荣获当代中国百名建筑师荣誉，2012 年

大量且多样的设计项目。这一时期的代表作品“雨山湖公园小品建筑”自由地运用现代主义建筑语言，并借鉴了一些中国古典园林元素，其形式克制内敛，但是一些元素有很强的修辞性，具有手法主义（mannerism）的特征。“文革”结束后，项秉仁重新回到南京工学院攻读硕士学位，并在 1981 年成为童寯先生的博士研究生。[3]两年后童寯先生仙逝，项秉仁遂转入齐康教授门下继续攻读，于 1985 年以《城镇建筑学基础理论研究》一文，通过博士论文答辩，成为中国首位建筑学博士。读博期间的项秉仁出版了“国外著名建筑师丛书”之一的《赖特》，并翻译了凯文 · 林奇的《城市的印象》。

在 1980 年代，西方现代主义建筑学内部多种诉求间的矛盾已经不可调和，符号学、类型学、形态学与文化人类学等思想方法已经兴起，形式与功能、符号与意义、结构与表面之间的关系已经脱离。这给项秉仁带来了莫大的困惑，也促使他深入研究现代主义本身的丰富内涵。他对赖特（Frank Lloyd Wright）与“城镇建筑学”的研究就是这些困惑的反应。从现在的眼光来看，研究生时期的项秉仁俨然是一位青年实验建筑师，他大量的投稿设计竞赛并获得奖项（剧场、车站等），将符号学、环境心理学等研究结合于设计探索，也参与、主持了一些重要的设计实践（侵华日军南京大屠杀遇难同胞纪念馆、马鞍山富园贸易市场、昆山鹿苑市场、杭州胡庆余堂药业旅游中心等）。

从 1980 年代中期开始，随着第一代建筑师退出日常教学，“Parti”与“Poche”这类“布扎”体系的术语渐渐不为人所知。东南大学在 20 世纪 80 年代中与瑞士苏黎世联邦高工建立了学术交流关系，在日渐频繁的青年教师交流中，建筑师赫斯里（Bernhard Hoesli）的空间与克莱默（Herbert Kramel）的构造思想开始影响东南大学的建筑教育。自此，从具体的教学方法上来说，“布扎”体系已经寿终正寝。

但是应该看到，“布扎”体系对中国建筑学教育的影响不仅限于其设计训练，也在于其隐含的职业规范，即建筑设计的运作管理制度。布扎的“图房”“草图”与“评图”等训练法本身是在西方传统的职业建筑师制度下建立起来的，将布扎体系完整地移植到中国并加以改造的第一代现代建筑师在任教授前，都有以职业建筑师身份进行独立实践的经历，这种气质也融入了建筑学教育中。建筑学的“职业主义”（professionalism）本应通过规训式（discipline）的教育方法完成代际传递并逐步合法化自身的正统性，但是这一规训方法不可避免地在社会主义阵营的设计院制度中遭遇尴尬，因为后者在很长一段时间压抑了独立自主的建筑创作。1984 年，王天锡开始主持新中国第一家建筑设计事务所——北京建筑设计事务所。1995 年注册建筑师制度建立，独立的“工作室”大量涌现。在前数十年在“图房”训练法中被激发的，但又在设计院制度中被压抑的个人创作欲望在整个1990年代中后期通过个人工作室的建立倾泻而出。由此，一方面，“图房”制度所暗示的建筑学自主性对项秉仁先生产生了潜移默化的影响，这些影响所累积的动力只有一个出口，即建立西方意义上的独立建筑师事务所。另一方面，“图房”对建筑学教育与实践的影响也设定了一种评价建筑师的标准，即在整个“后布扎”时代，即便建筑师所受的教育与训练方法开始逐渐摒弃草图与渲染训练，快速准确的设计表达能力依然是一个合格建筑师的最基本的修养。通过项秉仁先生的言传身教，学院派体系也跨越一代，潜移默化地影响到他的学生们，这一辈建筑师已经在各自的岗位上崭露头角，从他们身上依然可以观察到某种源于学院派的共性——对建筑学自主性的信仰，对形式伦理的高度敏感，对晚期现代主义语言的推崇，以及对手法与修辞的依赖。

## 第二阶段

### 缺席“实验建筑”

从1985年到1992年，项秉仁完成了从具有先锋色彩的青年建筑师向职业建筑师的转变。探索职业化的建筑设计道路一直是项秉仁的理想。但是当时的中国并没有单纯由建筑师构成的事务所，也没有注册建筑师制度，大型设计院里的“建筑专业人员”并不是西方意义上的职业建筑师。受限于事业单位的体制约束，设计院内的“建筑专业人员”无论在待遇收入与创作的自由度上都无法与建筑师事务所的同类人员相比。由此，项秉仁试图走出国门去探索建筑设计职业化的可能。就在项秉仁在境外进行职业化的建筑设计探索时，以张永和、马清运为代表的改革开放后第一代留学生开始回国参与实践，他们主张独立于体制内大型设计院或美国式大型事务所的第三条道路，他们的早期实践被称为“实验建筑”[4]。到美国以后，由于体验了建筑师事务所需要面对严峻的市场压力，项秉仁开始把注意力从前卫的建筑思潮转移到建筑师与客户之间的关系上。

在攻读博士期间，项秉仁的兴趣主要在后现代主义的各种理论。早在1984年，他就发表了《语言、建筑与符号》，这是国内最早介绍建筑符号学的论文。1985年，项秉仁获得博士学位后回到上海，进入同济大学任教。这时他已经完成了以富园贸易市场为代表的多个设计项目，这些项目本身是对当时的碰撞的建筑思潮的反应，是现代主义的功能体系与丰富的符号意义表达的结合体。1986年，顾孟潮、王明贤、赵冰等人和一些中青年建筑学人一同成立了“当代建筑文化沙龙”。沙龙的早期成员包括布正伟、马国馨、程泰宁、李大夏、萧默等，项秉仁先生也是成员之一。这一沙龙的主要议题是当时方兴未艾的“后现代主义”。王明贤在后来的《中国当代建筑文化思潮30年》一文中也回顾了项秉仁在富园贸易市场中的采用的“语义裂变”的设计方法，即对传统的四柱三间三楼的原型进行了戏谑式的改造。通过沙龙的讨论，项秉仁探讨将符号学和拓扑变换应用于江南水乡城镇的片段改造，引起了当时的先锋建筑师圈内的广泛回响。随后，项秉仁还参与了高名潞策划的“89现代艺术大展”。这是一个回顾整个八十年代中国艺术界心态的总结，同时它似乎也预示着一个时代的终结。

1989年，项秉仁选择了出国深造，先于美国亚利桑那州立大学任访问学者，并任城市设计研究组副导师。这期间他访问了赖特的西塔里埃森设计营与保罗·索莱里（Paolo Soleri）的乌托邦实验“阿科桑蒂”。随后，项秉仁先后在旧金山布朗·鲍特温事务所和TEAM 7国际设计公司事务所参与当地的设计实践，1992年开始任TEAM 7联营董事。这一时期项秉仁对建筑设计这一职业行为有了全新的认识，在一个完全市场化的建筑设计行业中，职业建筑师在为业主争取利益的同时更应当坚持伦理与操守。美国建筑设计事务所的“服务”式运作方式慢慢融入项秉仁之后的设计运作机制中。

1990年代初的西方建筑学已经分裂为日益繁杂的流派，精英的建筑学与大众的建筑学的分野也日益明显，占据学院主流的解构主义开始拥抱电子与媒介构成的知觉片段，这种趋势与项秉仁试图探索的建筑学职业化路途愈行愈远。美国时期的项秉仁虽然离“后现代主义”发生的现场更近，但是他发现这股思潮并不能撼动庞大且成熟的美国职业建筑师设计市场，因此没有继续符号学等“前沿”理论话语的探究。他搁置了“现代主义—后现代主义”的意识形态争辩，将前一阶段的学术探索转移到对建筑师的职业素养的反思中去，这种转变并非毫无前兆，也并未把符号学作简单化和庸俗化的理解。他对建筑符号学的探索是对现代主义的冷漠表情的修正，用符号来优化空间的感知与表义的传达，以提升建筑设计的质量与体验，然而他对现代主义建筑的基本信仰并未动摇。

< 于赖特设计的流水别墅前，1990年

> 访问伊利诺伊理工大学，1990年

1992年，项秉仁获得贝聿铭资助的旅美奖学金，并在完成旅行后与贝聿铭进行了一次谈话。随后听从贝聿铭的建议，奔赴即将回归的中国香港继续职业化的建筑实践。次年，张永和成立了“非常建筑”事务所，刘家琨弃笔重拾图板。1996年，王天锡的“北京建筑设计事务所”停止运营，“实验建筑”正式登上历史舞台，而1980年代的先锋——“符号学”却成为这一代新先锋的学术论战的标靶。曾经被认为可以借由重构现代中国精神的西方思想正在成为祛魅的对象，“符号学”是当时西方的后现代主义思潮的代表，也被视为抱残守缺式的建筑穿衣戴帽运动的同义词。实验建筑对“符号学”的指责反映了当时的年轻建筑师群体普遍对强势的形式主义的抵抗。事实上，“实验建筑”所指责的庸俗化的符号学与项秉仁研究符号学的初衷也是南辕北辙。符号学的庸俗化的主要推手是由于土地市场松绑所引发的整个建筑设计的急速市场化。而1980年代的“符号学”探索是为了丰富建筑设计的内容与语言，平衡现代主义对除功能以外的设计要求的忽视。当时还没有发生建筑设计商业化的趋势。1990年代的“实验建筑”见证了新一代建筑师的崛起，而此时的项秉仁已经远离了这种学术上的论争。

项秉仁无意中“缺席”了1990年代的“实验建筑”大潮，但是并没有停止对更纯粹的建筑语言的探索，后来的两个未建成项目——水清木华小区与建川博物馆，就代表了这种探索。随后，进入新千年的中国建筑设计市场经历着再一次的转型，虽然快速的城市化进程将建筑师引入一个黄金时期，但是建筑设计日益成为一种大批量的生产活动，建筑设计行业不断扩张，“时髦”形式的消费与再生不断加速，设计劳动的价值不断稀释，某种生产机制上的隐忧已经慢慢浮现。在新千年即将到来，这种隐忧尚未转化为危机之前，项秉仁创立了上海秉仁建筑师事务所。

< 于密斯设计的巴塞隆那德国馆前

> 于柯布西埃设计的马赛公寓前

缺席了“实验建筑”运动反而使得项秉仁能够将南京工学院所接受的全面系统的建筑学教育转化为一种职业建筑师的修养。项秉仁与1990年代“实验建筑”运动都在有意无意地抵抗1990年代的西方建筑学对现代主义理想的淡漠与对空间问题的疏离，只是“实验”建筑师的早期教育中已经几乎没有学院派的色彩，在面对强势的西方流行理论时也就会更自然地向经历过现代主义现场的第二代建筑师（黄作燊、冯纪忠、王大闳等）去寻找理论根源。而项秉仁则将早期对符号学的探索转换为对建筑设计的职业伦理的追索。与实验建筑特意强调对形式主义的抵抗不同，对项秉仁来说，形式、空间与社会服务都是内在于职业伦理的，强调建筑学的丰富内涵中的某一部分，以此来削弱其他部分的重要性是毫无根据的。这种对建筑学丰富内涵的全面把握客观上强化了建筑学的门槛与专业性，在为建筑师设立更高标准的同时，也在回应不断来自甲方与社会的新需求。

## 第三阶段

### “新现代主义”的显形

1993年到1999年，项秉仁任香港贝斯建筑设计公司总经理。此时他的建筑设计思想更趋成熟，前一阶段在美国习得的职业伦理使他几乎毫无障碍地适应了香港的设计市场环境，而他也同一些正在创业的企业家建立起合作关系。相比于美国，香港的建筑设计领域更强调服务业主的职业伦理，对空间的效率与价值的挖掘达到极致，同时会抑制一些相对个人化的形式表达。香港本土的现代主义具有西方1980年代涌现的“新现代主义”的一些特征，建筑评论家保罗·戈德伯格（Paul Goldberger）认为“新现代主义”是审美而非伦理意义上的，它试图重现现代

> 于赖特设计的罗宾住宅前

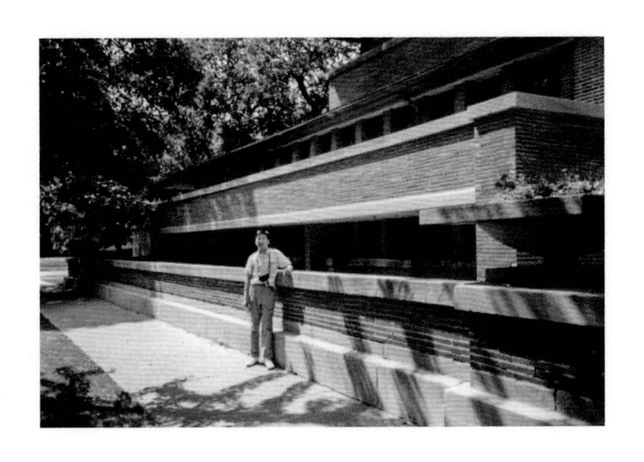

建筑的面貌并融入其他非现代主义的元素。在香港，“新现代主义”的设计态度已经融入地方的个性，它与20世纪中期的盛期现代主义的关系超越了风格上的承续。表面上来看，“新现代主义”通过将现代主义的一些形式语言符号化、修辞化来修正现代主义的功能主义、极简主义倾向，以回应当代社会的复杂文化环境；从深层探究香港的都市职业人士所推崇的精确、实用与沉稳的性格与“新现代主义”的气质比较协调，“新现代主义”自身对自由多元的形式语素的包容也能满足这个国际都会对多样性的需求。

香港在中国现代化的历史上一直担负着中介与桥梁的作用，在内地，复杂的政治运动造成现代化进程停滞时，在“港督”麦理浩所谓的“善治”下，香港恰巧完成了向一个制度完善、治理高效的大都市的转变。香港在经济上的成功迅速转换为文化上的输出优势。香港的城市建设也自然成为改革开放早期的内地学习的范本。1980年代开始，王欧阳事务所、吕元祥事务所、许李严事务所等香港的本土建筑师群体逐渐崛起，改变了巴马丹拿（P&T）、利安（Leigh & Orange）和马海国际（Spence Robinson）等老牌事务所几乎垄断香港的高端建筑设计项目的历史，他们的成就激励了其他华人/华裔建筑师。在建筑设计理念上，香港几乎没有受到后现代主义的影响，本地特有的实用主义导向消解了各种“前沿”西方建筑学思潮的冲击。因此，现代主义建筑学的功能主义理念似乎在采用自由放任经济的香港得到更彻底的推行，这种功能主义与大企业的价值观与审美偏好融合在一起，成为当时迫切寻找现代化范本的中国内地建筑师的习仿对象。

1990年代正好是新现代主义建筑在香港开始涌现的时期，在《当代香港新建筑观察》一文中，项秉仁提到了几个“新现代主义”的优秀范例，包括严迅奇设计的香港大学的研究生楼、香港国际机场港龙航空暨中国民航（集团）大楼、英国驻香港领事馆、香港历史博物馆，等等。这些作品都具有手法主义色彩，但不背离现代主义的形式语库与设计范式。在香港时期的项秉仁承接了大量的香港与内地其他城市的设计项目，其中包括较多的办公、酒店、住宅与一些室内设计项目。“新现代主义”比较符合大型企业的总部办公的公共形象与气质。这个时期项秉仁完成的主要设计有湖南国际金融中心大楼（1995）、武汉东方风情俱乐部（1995）、龙柏苑住宅小区（1996）、深圳德港公司室内设计（1997）、常德市芷园宾馆（1998）、江苏电信大楼（1998）、南京电视电话综合楼（1998）、南京泰山新村长途汇接楼（1999）。项秉仁对不同的项目类型会采用相适应的设计方法，其中大型企业的总部大楼（如南京的几个项目）比较完整地表达了西方晚期现代主义的构图、语言与色彩喜好，其中泰山新村长途汇接楼还可以看出当时新落成的香港九龙塘又一城购物中心立面的影响。而其他的一些酒店与住宅项目（如武汉东方风情俱乐部）则在功能化的平面上采用不同的风格元素，试图对建筑文化传统进行当代语境的回应。无论采用何种风格，项秉仁对形式语言的选择是极其慎重的，他尽量使用点、线、面构成的基本几何形来表达高尚的审美趣味与复杂的文化隐喻，这依然隐约可以看到早期符号学研究的痕迹。

1999年，项秉仁回到同济大学任教授暨博士研究生导师，担任建筑设计方法论团队的责任教授，并成立项秉仁建筑设计咨询（上海）有限公司。在随后两年，他完成了光明邨酒店立面改建、复兴公园大门、杭州富春山居、卡子门广场标志物、广州国际会展中心与北京首都博物馆竞赛方案。这些设计作品并不具有统一的风格，每一个作品都在回应各自所处的社会场景与文化语境，都具有独立且完善的形式语言系统。上海复兴公园大门是这一过渡时期的代表，这是一个城市的小品建筑，然而却因为与项秉仁早年经历的渊源，受到了极为特殊的对待。为了复现法国式的木门，项秉仁找到了原大门的档案图，辅之以素混凝土与钢栏杆的现代手法。为了设计门头，项秉仁邀请了罗小未先生担任顾问，最后，虽然大门门头依然跳脱不出西方古典的形式构图，却是借用现代的点线面形式语言来实现。

2001年，当时的同济大学青年教师王方戟对项秉仁进行了一次采访，采访内容被整理为《观察与思考——访项秉仁建筑师》一文。文中项秉仁明确表达了现代主义的设计原则与适度的自由之间的关系，并第一次将自己的设计思想归纳为"新现代"。在谈及"现代"在当代中国的含义时，项秉仁认为现代主义建筑思想的本质是形式由逻辑推理来决定，解决问题本身就是设计目的的一部分。对理性的坚定信仰是项秉仁的设计思想的出发点，理性是探究对象本质的重要方法，这并不等同为功能主义，而更接近自启蒙时代以来的人本主义理性。在化身于20世纪的盛期现代主义建筑之外，在不同的语境下，理性可能表现为不同的样貌。对于迪朗（Jean-Nicolas-Louis Durand）来说，理性可能是根据空间部件来排列整体布局的经济与高效，对勒杜（Eugène Viollet-le-Duc）来说，理性可能是材料与工艺所限定的建造规律，对翁格斯（Oswald Mathias Ungers）来说，理性可能是自主、永恒的形式准则。项秉仁的设计理性可能包含了以上所有的方面，它更是介于规则与自由之间的伦理标尺。

在建筑设计事业进入成熟期后，新千年后的项秉仁开始探索人类进入数字化时代后的建筑学实践方法，在21世纪初网络技术的飞速发展的影响下，不少嗅觉灵敏的建筑师和学者对于网络时代的城市和建筑的发展趋势曾掀起过一阵探索热潮，恰逢学院要求各学科组确定研究方向，项秉仁作为责任教授便根据教学小组的研究兴趣，确定了数字化建筑作为研究专题之一，也在日常教学与研究生培养中强调数字化时代的各种技术、社会与文化条件对建筑学本体的影响。在《面对数字化时代的建筑学思考》（2001）一文中，项秉仁认为建筑师的传统工作模式已经开始面临电子商务、社交媒介、虚拟现实与新兴的计算机辅助设计与建造工具的挑战，建筑学的精神与物质内涵也在经历深刻的变革。但是，即使建筑学是一门非常具有社会性和实践性的专业，建筑师对社会问题的关注和了解最终将落实到具体物质环境的营造，而物质环境的基本构成要素依然是空间和形体。数字化时代的洪水猛兽无法动摇项秉仁对建筑学本体的根本认识，当然建筑学本身也必须回应这种"时代精神"，这种处境与项秉仁在1980年代所面对的后现代主义符号学冲击并没有根本差异，数字技术是实现新条件下的理性的工具，理性依然是驱动建筑学发展的动力。

研究上的深入促成了项秉仁的"新现代主义"语言的丰富。在合肥文化艺术中心的设计中，他就采用了双曲面的建筑形态，并通过化整为零的手段使建筑更有肌理和丰富的表情，具有强烈的动态性，并因这个项目获得了中国建筑艺术奖。虽然采用了大量类似母线生成的曲面，这个作品的风格依然不脱离"新现代主义"的基本框架。同样，宁波东部新城行政中心更类似迪朗或翁格斯的形式排列组合，不对称的形式与架空的底层消解了行政建筑的压迫感，也展现了开放的姿态，这是对理想的行政管理形式的一种期许。这种如同乐高玩具一般的形式组合游戏反映了一种自主的建筑学，即形式可以借由一种预先设定的语法自行生成。同一区域的宁波文化广场是一个容纳了科技馆、电影院、儿童活动中心、妇女活动中心、健身中心的综合城市功能区，项秉仁用规划开放式街区的方式预先对整个项目进行城市设计导则的编制，再发动设计团队的建筑师各自设计一个街区，以模拟自发式造城的过程。

在大唐不夜城文化交流中心的设计中，项秉仁采用了一种更具挑战性与颠覆性的设计策略，即用现代主义的基座与唐风建筑大屋顶的分离来应对风貌、功能与环境的不同要求。项目所在的大雁塔广场的南侧已经有一些仿古建筑，但是这种相对粗糙的简化反而损伤了广场本身的严肃性。项秉仁认为，大唐不夜城文化交流中心处在西安的旅游区内，业主也有一定的投资来改善整个区域的形象，大众的体验是项目的首要考虑，旅游建筑本身的功能要求应该被尊重，形式不必受限于伦理维度。这种操作在项秉仁的以往实践中是较少见的。当然，唐风建筑在国内目前的建筑实践中大量存在，具有一般的仿古建筑所无法企及的文化与政治的合法性基础。唐风本身经过几代建筑师的不断考据与诠释，已经发展为可以再创造的形式语法，与西方新古典主义时代的建筑语言遥相呼应。项秉仁试图在细部上复现唐风建筑的屋顶形式，以回应整个城市的整体环境要求。原真的唐风屋顶与功能主义的基座的并置并不突兀，相反却保留了各自的独立性。

回沪后的12年间是项秉仁建筑创作的高产期。以合肥大剧院、宁波东部新城行政中心、大唐不夜城文化中心三个项目为代表，新千年时期项秉仁的创作已经完全跳出了各种主义与学理的羁绊，以服务业主、满足使用者预期、平衡功能、实现高完成度的作品为首要的目标，在某种意义上，他的建筑创作获得了新的自主。这种自主表现在设计方法与工具的不拘一格上，也表现在形式、功能与意义的张力上。[5] 近期项秉仁先生将工作重心逐渐转移到培养并支持年轻一代建筑师上，自觉地减少在一线的建筑创作工作上的参与度，但是依然保持着对当代建筑学发展趋势的高度关注。项秉仁的"新现代主义"建筑观在新千年后开始针对社会语境发展新的内容。

项秉仁认为，新的建筑创作环境下，经典现代主义的“洁癖”已经不再适用，这反映了现代主义初生时拮据的经济条件，即将多余的装饰性构件视为某种违背伦理的罪恶。现代主义对功能的强调固然提供了切入建筑设计问题的方法，但是却抑制了文化与情感因素的表达。在新的经济条件下，应该允许建筑师在满足基本功能之外的形式探索。

## 结语

将项秉仁先生的人生经历分为三个阶段分别评述，固然有利于厘清项秉仁在不同历史时期根据社会与个人境遇所作的即时即刻的建筑观调整，有利于建立一种连续的个人叙事，却容易落入编年史的历史决定论窠臼，并忽略对不同历史阶段的经历进行交叉印证的可能。从接受学院派的建筑学教育，到树立现代主义建筑观，再到通过后现代主义的语言手段来丰富现代主义的表达，最后以一种“新现代主义”方法体系获得新的自主性，项秉仁的建筑观发展至少可以注解并例证中国当代建筑史中一些悬而未决的问题（在篇首这些问题已经提出）：

1. 传承并改造自“布扎”体系的学院派建筑学教育的遗产需要进一步评估。项秉仁所接受的早年教育代表了以完整职业能力训练为目标的培养体系，这一体系注重建筑师应对大型工程的能力，尤其强调将构图与造型的概念贯彻到底，建筑师必须对从设计到实施的整个过程负责。以往的建筑评论偏重对这套体系在学科本体领域的影响的分析，重点关照现代主义—新古典主义、功能—形式、政治性—自主性等对立概念的辨析，却忽略了它对职业观乃至职业伦理标准的设定与暗示。学院派的美术教育与综合的职业训练固然为中国的快速城市化培养了一批高素养的建筑师，也长期界定了建筑师在整个社会生产体系中的核心劳动价值。与此相对应的是，当下的建筑设计项目转向“小而专”，面向大型工程的综合能力不一定是必要的，全能型素养在人工智能、大数据与共享经济的大潮下正在经受不断的考验。项秉仁在学院派建筑教育中所接受的偏重学科本体的职业伦理在个人的设计实践中逐渐转化为服务社会的职业伦理，这也印证了这两种职业伦理在学院派教育中的一致性。

2. 自1980年代初开始，由于完整接受学院派体系的老一辈教育家相继退出本科教学，以柯林 · 罗（Colin Rowe）与赫斯里（Bernhard Hoesli）等晚期现代主义建筑学家为代表的、具有较高操作性的建筑学设计方法填补了建筑学教育的方法论“空窗期”。同时，符号学、类型学、文化人类学的一波波冲击相继到来。在这样一个激荡的时代，由于以大设计院式的生产为导向的建筑学价值观的巩固与延续，所有的思想激流都无法逃避快速“变现”为操作方法的命运。那些抵抗生产体制的建筑师则汇聚为一股潮流，“实验建筑”正因糅合了“操作”与“抵抗”的集体诉求而被指认为历史事实，“实验建筑”本身是在先行一步的西方晚期现代主义建筑学的话语结构下对中国当代建筑史所作的关照，它将建筑设计所面对的背景抽象为一些关键约束条件，使得建筑学更加“纯粹”，“实验建筑”在无意识中接受了“抵抗”所虚设的背景，却消弭了建筑师这一行业在中国快速城市化进程中的职业行为的复杂性，降低了与技术、权力与资本进行议价的难度，并扩大了进行形式操作的舒适区。在“抵抗”与“操作”所构成的维度上，不同建筑师会有偏向。建筑学的自主性以何为据？对复杂社会需求的迎合究竟可以到什么限度？观察在“抵抗”与“操作”中分化出的不同态度，成为分析当代建筑师群体的方法。那些完整接受学院派教育的建筑师，往往会在这个天平上选择偏向可操作性的那一端。虽然项秉仁并没有直接卷入“实验建筑”的潮流，但是这一维度依然是理解他的建筑观的工具。[6]

3. 从美国（现代化西方的代表）与香港（现代化东方的代表）汲取设计方法论养料是当代建筑师的普遍经历。美国与香港的建筑师制度决定了对业主和受众的“服务”势必超过建筑学自身的学科要求，建筑作品首先必须是一种易于读解、激发愉悦的空间对象，然后才是建筑师观念表达的载体。这背后的假设是，存在一种跨文化超地域的全新现代主义建筑体系，这一体系已经摆脱了20世纪中期的盛期现代主义的一些偏执与洁癖，获得了更丰富完整的内涵。而位于全球化风口浪尖的亚太地区则是孕育这种新现代主义的沃土。香港是项秉仁的建筑之旅的最后一个驿站，在香港之后，他的“新现代”建筑观全面显形，并非巧合。新千年中，项秉仁的建筑创作更趋自由，更大胆地运用形式语素，但仍不失为他的建筑理念的合理发展，也是早期学院派教育的灵活性与适应性的体现。

本文试图在个人历史叙事、学科历史叙事与社会历史叙事的多重线索中穿行并保持平衡。项秉仁先生的事业与经历表现了改革开放后一代建筑师的共性，比如早期接受严苛的学院派建筑学教育；始终如一的修正、发展并丰富现代主义建筑的内涵与语言；学术上的探索同步于工程实践活动；具有跨文化的专业与文化视野（中国内地、美国、中国香港）；兼具深厚的学术与实践背景；建筑教育思想与实践一脉相承；致力于现代建筑师事务所制度的确立，等等。项秉仁的“新现代主义”建筑观的成熟期在新千年以后。这一观念体系的成熟不仅直接铺垫于其个人的坚实的建筑学教育，得益于亦波折亦丰富的人生经历，也与他个人事业的运营方式的完善不可分割。因此，对项秉仁的建筑理念发展的解读，既可以为年轻一代建筑师呈现可资学习的范本，也能够为当代中国建筑史的书写构建可供品鉴的案例。这一解读将当代中国建筑学发展的大叙事拉回到建筑师执业现场与建筑学话语变迁的具体语境中，以期在激流中寻找建筑学自我鞭策的驱动力。

## 注释：

1 第二代建筑师在1940年前后接受教育，多少都通过实践、媒体与海外留学等途径，接触了当时已经成为主流的现代主义建筑学，并在整个1940年代到1950年代初在教学中鼓励实用与创新的思维，这在圣约翰大学、之江大学乃至中央大学的教案中有所表现。1952年后，随着苏联教案的引入，学院派教育重新成为国家意志，因此，1949年后的第三代建筑师多数都是在严谨的学院派训练中成长起来的。“新三届”（1977、1978、1979级大学生）受教育早期，学院派的影响还未式微，随后，各个主要建筑院校开始转型，西方后期现代主义思想不断冲击既有的学院派体系。张永和1977年春考入南京工学院建筑系，1981年赴美留学，是“新三届”中最早具有国际思维和融入西方建筑文化的成员。项秉仁的学习历程横跨第三代建筑师与新三届，属于最后完整接受学院派教育的一代。

2 王文卿和吴家骅认为，南京工学院的基础教学从20世纪40年代到80年代经历了三个阶段，分别以“西古”“中古”和“现代”的渲染练习为代表，详见顾大庆：《中国的“鲍扎”建筑教育之历史沿革——移植、本土化和抵抗》，《建筑师》第126期。

3 当时全国仅有五位博士生导师——同济的冯纪忠，东南的童寯、杨廷宝，华南理工的龙庆忠，清华的吴良镛。硕士毕业生中符合条件且愿意攻读博士的也非常少。

4 “实验建筑”这个词可以追溯到1996年于广州华南理工大学召开的“南北对话：5·18中国青年建筑师、艺术家学术讨论会”，最初由王明贤与饶小军提出。饶小军曾撰文提到这次会议明显带有“实验性”和“前卫性”，并有强烈的“反理论”色彩（当时主导的西方理论），会议涉及中国的“实验建筑”的可能性和未来发展路向问题。史建认为“实验建筑”终结于2003年，那一年所举行的“非常建筑，非常十年”回顾展和研讨会，被看作“实验建筑”的终结和“当代建筑”的起始。

5 对于新现代主义设计实践，项秉仁提出六点理性原则：①要满足甚至高于业主期待；②从建筑所在地区或城市文化层面出发构思；③自觉考虑整体城市设计的目标，包括建筑形体和城市空间；④满足功能预期；⑤尽量运用当代最新建筑技术和材料并在设计建构中体现出来；⑥注重建筑细部的刻画和高完成度。

6 项秉仁认为建筑学学科的发展主流还是应该切合时代科技和社会发展潮流，而“实验建筑”是一个包容各种探索的集合体，其中既有顺应时代发展的尝试，也有以个人趣味为导向的个性创作，不能一概而论。

## 参考文献：


[1] Baird, George. OMA, “Neo-Modern,” and Modernity: George Baird in “Conversation” with the Editors of Perspecta 32[J]. Perspecta, Vol. 32, Resurfacing Modernism (2001): 28-37.

[2] Goldberger, Paul. Architecture View; Modernism Reaffirms its Power[J]. The New York Times, November 24, 1985.

[3] 顾大庆. 中国的“鲍扎”建筑教育之历史沿革——移植、本土化和抵抗[J]. 建筑师，2007(4).

[4] 顾大庆. 中国建筑教育的历史沿革及基本特点[M]// 朱剑飞. 中国建筑60年（1949-2009）. 北京：中国建筑工业出版社，2009.

[5] 钱锋. 中国现代建筑教育史(1920~1980) [M]. 北京：中国建筑工业出版社，2008.

[6] 饶小军. 实验建筑：一种观念性的探索[J]. 时代建筑，2000(2):12-15.

[7] 史建. 从“实验建筑”到“当代建筑”[J]. 新观察（第一辑）.

[8] 王方戟. 观察与思考——访项秉仁建筑师[J]. 时代建筑，2001(1):42-45.

[9] 王建国，单踊，刘博敏，等. 转折年代“中国现代建筑教育摇篮”的继承者与开拓者们——以东南大学建筑学院“新三届”学生发展研究为例[J]. 时代建筑，2015(1):18-25.

[10] 王衍. 从“异端”到“异化”：深圳城市化进程中的社会条件和60后建筑师实践状况[J]. 时代建筑，2013(1): 46-51.

[11] 朱剑飞. 中国建筑60年（1949-2009）：历史理论研究[M]. 北京：中国建筑工业出版社，2009:192-200.

[12] 周卜颐. 建筑教育的改革势在必行[J]. 建筑学报，1984(4):16-21,52.

[13] 周庆华.“实验建筑”？“当代建筑”？——思考中国当代实验性建筑[J]. 新观察（第六辑）.

华 ..... 华霞虹
项 ..... 项秉仁
滕 ..... 滕露莹

# 时代与个人的选择

## 华霞虹访谈项秉仁

# The Choice of Times and Self

## Hua Xiahong Interviews Xiang Bingren

华霞虹，同济大学建筑与城市规划学院教授、博士生导师。

访谈时间：2018 年 11 月 3 日，11 月 15 日，11 月 23 日
参与者：滕露莹、晁艳、周希冉、吴皎、朱欣雨
访谈地点：上海秉仁建筑师事务所会议室

## 录取南京工学院建筑系

华　项老师，当时您考大学第一志愿就想去南京工学院吗？

项　不是，我一开始没有想要读建筑，而是想先报考艺术类院校——北京电影学院，学电影摄影。因为我从小就比较喜欢画画，希望今后自己的职业能与兴趣相结合，电影摄影就是一个技术和艺术相结合的专业。初试时200人里挑20个，我选上了；复试再从20个里挑出2个人，要面试，结果我没考上就很沮丧，连电影院都怕去了。

我有一个很好的班主任老师，会吹奏长笛，他建议我填报南京工学院建筑系。一方面当时受到意识形态的引导，认为应该志在四方，到祖国需要的地方去，到祖国需要发展的领域学习；另一方面考艺术类院校受挫后，对于报考当时录取分数线较高的上海高校有些犹豫，于是第一志愿选了清华大学的农业科技专业，南工只是第九志愿。可能南工当时有老师到我的高中——向明中学来选学生，我的班主任老师觉得我适合学建筑学，做了推荐。但是我拿到录取通知书时并不太高兴，因为是第九志愿。

我后来反省，觉得自己在高中的时候，搞了很多课余活动，不像其他同学在高三那一年拼命专注于学习。结果他们考上了上海交大、复旦那些本地学校，而我就无奈退居其次，看来人读书是要拼一下的。所以到了南京工学院以后，我觉得要老老实实读书了。

还有一些有趣的孩子气原因。1961年刚进大学开新生同学会的时候听到一些言论，他们说进大学后有一个现象，像上海这种大城市来的同学，中学普遍比较好，所以开始的时候学习是领先的，但后来就逐渐落后，到毕业时就很普通了。我不服气，就暗下决心，不能让这种传言成为现实。于是我在南工念书很卖力，一年级到五年级所有的考试成绩都是五分，当时在南工有些名声。其实建筑学这个东西也很难讲，真正数理化很好的人也不见得就一定学得好，还是契合自己的兴趣，才能学得好。

华　当时您在南工本科学习的时候，是像杨廷宝、童寯、刘敦桢先生这样的中国第一代建筑师在执教吧？您感觉如何？

项　是的。当时是非常崇拜的。因为无论从学问，到手头功夫、画画，或者是待人接物、仪表方面都觉得我们这代接不上了，他们就是我们的榜样，肃然起敬。当时童寯先生基本上是搞研究，指导那些老师，跟学生已经距离很远了。刘敦桢先生也是教老师，基本上不教我们低年级的。但杨廷宝先生还教我们建筑初步，他在黑板上拿粉笔画的柱头，轮廓线条画得非常准确到位，令人折服。

华　同济建筑系在20世纪50年代末60年代初，因为“教育革命”要多招收工农出身的学生，培养成为一竿子到底的全面手，也引起了很大的矛盾。比如谭垣先生，他们觉得建筑师需要有专门的艺术文化修养，而学生则会批评这是资产阶级生活方式。当时的阶级差异和矛盾还是很明显的。

项　是的，当时学生的背景的确很不一样。有的学生从苏北农村来到南京这个大城市，在花花世界禁不住诱惑，就会出去吹个头发，买个奶油蛋糕之类的。我在班里一直担任班长，当时就批评他们，你爸妈现在正在外面逃荒讨饭，你却在这里挥霍，乱花钱。

华　20世纪60年代初，南工的教学系统有做人民公社规划这样的新类型吗？还是一个比较传统的类型化的设计方式？有没有在设计院实习？

项　教学上基本上是类型化的，但是到毕业设计的时候，类型会多样一些，比如说工人居住区改造，或者是像你讲的人民公社规划。有设计院实习，差不多一个学期。

华　是以真题为主，有去画施工图吗？比如说跟外地的设计院去画施工图。

项　那是实习的时候了。毕业设计基本上还是方案设计，毕业设计比较主张实际的项目，会去做调研，但是不一定做到施工图的深度。

华　平时一个学期大概做几个设计？

项　一个学期一般做两个设计，从小住宅到火车站、剧院这样的类型设计。我学生时期设计的一个火车站还曾刊登在1964年第五期的《建筑学报》上。
我1966年大学毕业后，报考了杨廷宝的硕士研究生，应该是录取了。当初导师很少，有资格带设计研究生的就是杨廷宝先生。因为我当时成绩好是全系有名的，系里一些老师鼓励我去报名，但后来国家把研究生取消了，所以没有念成。

## 研究美国建筑师赖特

华　项老师您"文革"结束后回南工读研，马鞍山的设计院对您有限制吗？还是会鼓励您回学校读研？考研有困难吗？

项　当时是整个国家大的形势，政府鼓励大家回去读研究生，所以一般单位不太敢限制你。因为我1966年毕业，1967年走上工作岗位，到1977年恢复高考，已经工作了10年。我不知道能不能考得上，也不知道学校里老师要不要我，就先到学校去打听，找了齐康老师。齐康老师说了一句话，让我一下子信心大增。他说："你要回来，不考我也要你。"后来考试考得也挺好，主要考了英语、政治，还有一个快题设计。

华　当时有多少人去考？

项　当时考的人也不多。因为第一年恢复研究生制度，很多人也不愿意考。其实我本来也不愿意考了。因为我们南工建筑系有一位教授，是美国回来，叫成竞志。他是在美国事务所工作很多年后，大概1950年代回国的。回来以后因为家庭出身不好，在"文化大革命"中受到很大的冲击。他就觉得回来回错了，回学校也回错了。在我要考研究生之前；我去问他该不该考研究生回学校？他说你现在工作好吗？我说我现在工作很好，在设计院做建筑设计。他说你生活条件好吗？我说也可以，有一套房子。他说："我看你还是算了，别考了。"因为他实在太伤心了。但我最后还是去考了。

华　当时南工有没有单独的导师指导？20世纪70年代末80年代初，同济建筑系这边是有专门的导师的，比如像吴景祥、王吉螽、黄家骅这些老师都是独立辅导研究生的。听说南工是以导师组的方式指导研究生的，是吗？

项　我们是三位指导老师，刘光华、齐康和钟训正，什么原因，我也搞不清。实际南工第一届建筑设计就招了三个研究生：我、仲德崑和黎志涛。黎志涛比我大一点，他是清华六年制的。所以是三位老师，三个学生。我们读研究生的时候，正好张永和他们读本科在大教室里面上课，我们还作为老大哥去看他们。

华　中国建筑师对于中国建筑怎么实现现代化发展，始终有使命感的。在您开始读研究生做赖特研究的1980年代，建筑界更加积极地思考这个问题。当时好像几个主要的高校交流还蛮频繁的，比如您在论文中也提到，像罗小未先生也提供了资料。您能再介绍一下当时的学术状态吗？硕士论文您为什么选择研究美国建筑师赖特呢？

项　第一代建筑师给我们带来的（布扎系统的）建筑教育思想，是很复杂的一套东西，当时在南工根深蒂固，等于是历史传承。我们在坚持这套教学的时候，国际上实际已经远远抛弃了布扎的教育，都是采用现代主义的教学理论。不过当时国内在建筑领域，好像把资产阶级思想和现代主义划等号了，政治上比较怕资产阶级自由主义思想影响到我们，所以对于现代主义建筑思想的教育，一方面是抵制，另一方面也因为长期抵制的关系缺乏教材、缺乏教师，教师都不了解国外的发展情况。清华大学的吴焕加、南工的刘先觉，在教建筑历史时，基本上把现代主义这一块缩到很小的范围。现代主义的小住宅、开放空间、流通空间，都是跟人家的生活方式联系在一起的，所以他们特别紧张，特别怕学生受到影响。但是反过来不讲又不行，因为跟国际上的差距太大了，这些搞建筑历史和理论的老师也明白这些道理，所以不把它们作为建筑历史和理论课程的内容主体多讲，而是在后面讲当代西方的建筑发展时，用一种批判性的姿态去讲。所以很矛盾，但是我们学生都非常感兴趣。

华　这是在读研阶段，还是大学本科阶段？

项　是本科阶段，1961年到1966年间。那个时期，封闭的东西学生反而感兴趣。当初我们实际上也看腻了古典的那套东西，觉得是一成不变的，从文艺复兴到新古典主义的知识，用又用不到。对现代主义比较有兴趣，但是又被抑制住了，不给讲太多。所以最后在课堂上就知道了四位现代主义大师，勒·柯布西耶、赖特、格罗皮乌斯和密斯，介绍他们的一些设计作品和设计理论。当时非常震撼，因为看到建筑可以做得如此开放，能跟现代人的生活和现代的科学技术相结合，就对这个感兴趣。但是我们在本科阶段，这些知识都是像吃甜点一样给的，只给一点点，我们就没能学到什么东西。

毕业以后走上工作岗位，发现现代主义这套东西非常有用。因为第一是功能主义的理论，形式服从功能，就是说要把功能安排得最好，才能做出一个好的建筑；第二，建筑形式不是来自一个刻板的模子的重复，而是根据它的功能、环境来考虑。我们觉得这些观点很新，所以自然而然认为现代主义思想是科学的、进步的，我们应该更多地了解这些东西。

当时我觉得这四位大师中间，赖特的影响比较大，第一是他有非常好的作品，这些作品不但现代，而且结合了美国的历史和文化；第二是他的理论最多，如果研究赖特的话，相关的参考书也最多。

因为大学里没有好好学到现代主义，也没有好好了解现代主义大师的思想，所以就想在研究生阶段，把他的东西好好了解一下。另外一个考虑就是想把系图书馆中关于赖特的英文书啃一啃，一方面能够了解内容，一方面可以学学英语。

华　选择赖特做论文课题是您自己提出来的吗?

项　是的。提出来以后就征求导师意见,我当时的指导教师是刘光华。刘光华是哥伦比亚大学毕业的建筑师,跟贝聿铭是同龄人,今年100岁了。刘老师觉得挺好的。在硕士阶段我做了两件事情,一件事就是对赖特深入研究,当时国内对赖特的研究还是很少的,遗憾的是当时没有条件去美国。

华　现在您也是赖特研究的权威。后来接续的研究其实比较少,现在基本上都是直接引进书,包括他自己的自传。

项　硕士研究生阶段我做的第二件事是翻译了凯文·林奇的《城市的印象》。因为当时正好有一位澳大利亚的教授到南工来做讲座,提到这本书比较重要。其实我当时也不了解,那本书拿来一看很薄,就想把它翻译出来,没想到这本书在城市设计学科里是很经典的。

华　赖特研究和翻译《城市的印象》都非常具有开创性。当时在1980年代中国建筑界翻译出版了很多西方理论经典著作。您研究的两个,正好一个是建筑方向,一个是城市设计方向,都很经典。您是硕士论文写完之后才知道要编建筑大师专辑的吗?

项　对,我先写了论文,后来中国建筑工业出版社就定了要出赖特的专辑,要找人写。他们去找汪坦,他曾经在赖特事务所工作过一年。汪坦先生正好是我的硕士论文的评阅老师,他就推荐我了。

华　当时是不是因为各校研究生很少,所以像论文评阅、答辩,实际上很多是跨学校的?我看见您的博士答辩安排在一个阶梯教室,来了好多人,连冯纪忠先生都过去了。

项　是的,因为第一次,没有经验,导师和学校也没经验。我没看过一本国外的博士论文,当时一本也看不到。直到今天我们还在谈,建筑学究竟要不要博士。当时我读博士的时候就有这个观点,我们的导师很多都不是博士,杨廷宝、童寯先生都不是博士。

华　感觉做建筑设计硕士就够了。您很超前,因为当时没有什么人读博士。

项　不是我超前,国家搞了一个博士制度,要选几个老师出来当博士导师。我如果工作落实了,就可能不会去念了。人人都说童寯先生是学术泰斗,我觉得要念就念他的博士,他不要就算了。

华　所以实际上您是有点机缘巧合地去读了博士。我还想了解一下,研究赖特对于您认识建筑,或者说对于那个时期的中国建筑学界对建筑学的认识,有没有起到一定的作用?

项　赖特的作品,不一定所有人都喜欢,但是他的理论确实感觉是放之四海而皆准的。因为首先他反对古典主义只讲究形式不讲究功能的陈腐建筑观念;其次他提倡建筑要追随材料的本性,建筑要跟所处的环境有机结合,造的房子像从地上长出来的;再者,建筑要体现自己的本土文化,反对美国人盲目追随欧陆建筑样式。他的塔里埃森建筑营地着重培养学徒的文化艺术修养和建造技术的实际操作能力,力图改变传统建筑教学的弊病。

华　这些是跟您原来在设计院做设计时就认识到的现代功能主义原理一致的，还是一些新的启发?

项　我们这一代人经历了传统的古典主义的巴黎美院建筑教育的影响，也受到了现代主义建筑思想的冲击。实际上，现代主义在某些方面是符合我们国家五六十年代提出的“适用、经济、在可能条件下注意美观”的建筑方针的，所以在中国也得到了一定程度的推广。虽然没有打现代主义的旗号，但是在设计院的职业建筑师的创作方向或操作方法，还是受到现代主义很大的影响。现代主义后期受到了批判，因为它太国际化了，不考虑本土的环境，搞得千篇一律。赖特比较注重建筑物跟当地的文化、环境结合起来，要住美国人自己的房子。我们国家好几代建筑师也希望做到所设计的建筑不但是现代的，而且是中国的。比如1959年北京“十大建筑”，虽然是一个折中主义的东西，也是想走这条道路。

华　这些中国建筑某种程度上是非常符号化的，社会主义内容，民族主义形式。但是赖特比较讲究建筑跟基地的关系以及材料的运用，并不是一种符号化的方式。赖特的设计虽然有装饰，但是整体并不是一种形式符号化的创作，因此跟当时中国的状况有很大的不同。

项　是的，我们中国的建筑师都有这样的愿望，但是都还很肤浅，没有很好的作品。王澍其实也在做这件事情。凡是受过良好教育的建筑师，可能内心都有这种冲动和愿望，要做出既是现代的又是中国的作品。

刚才你讲关于中国建筑现代化的问题，中国建筑师对于我们国家的建筑和城市建设的现代化诉求，其实都很强烈，也很混乱。因为中国的建筑师群体始终挺复杂的，有老一辈从国外留学回来的，也有在国内自己成长起来的。有受了各项政治运动、各种思潮批判的，还有受到国外理论影响的。中国建筑师本身对于建筑的认识也是很多样化、很复杂的，所以很难概括“中国建筑”。因为中国是个太大的国家，而且又有比较长的发展过程。

我倒是觉得，建筑师或者建筑理论家把建筑从专业领域的角度看得比较重，愿意花很多时间去研究，也会提出各种各样的看法。但是建筑毕竟是一个受到社会、政治、经济发展影响的产物。我年轻的时候看到城市里冒了一个新建筑出来，会觉得这个新建筑造型很不好，跟原来的城市不协调；也会觉得做这个的建筑师很拙劣，没有专业操守，随便做个东西，然后就造起来放在城市里面。于是城市里老是有那些烂建筑，包括我们陆家嘴，一开始是国际会议中心的两个球，后来又出现平安保险的古典主义的大楼。大家都对此议论纷纷。

但是后来在国外旅行多了，著名的大城市都去过了，就觉得实际上每个城市都不是那么理想化的。没有一个人的理论和观念能彻底影响城市。建筑师的力量实际上很弱小，城市始终是受到政治、经济、社会文化发展的影响。所以反思我们建筑师是不是在寻求一种不可能达到的境界? 现在上海比以前好一点，专家的一些观念影响到领导，领导也知道“城市有温度，建筑可以阅读”这些观点了。但事实上建筑师的影响力是有限的。主宰一个城市的形象和面貌绝对不是建筑师的能力可达的。因此我有时也在想，我们建筑师是不是把自己看得太伟大了，把自己的作用看得太大了? 我们只不过是一个专业，而且这个专业也没有一个统一的标准，每个时期都有不同的理论、观点占上风，所以城市里难免会出现很多奇奇怪怪的、不协调的建筑。其实城市就是这样，我们做好自己专业的事情，也就不会耿耿于怀了。

## 中国第一位建筑学博士

华　后来您怎么读了童寯先生的博士呢?

项　说来话长。因为"文化大革命"没有读成研究生。毕业分配时，官方意志是要打破原来的资产阶级路线，把成分好的工农阶级出身的学生分到上海的大设计院；出身一般的或不好的，就分到相对差的地方。所以当时就把我分配到武汉中南电力设计院。去了那边以后又下放，大学生要到工厂农村去锻炼，所以又分到了鞍山钢铁公司，后来又回到马鞍山，在马鞍山建筑设计院工作了一段时间。后来国家恢复研究生考试了，才再考回了南京工学院的研究生。

研究生毕业之际，南京工学院想让我留下来教书，我推辞了，主要是我觉得作为一名建筑师应该造点房子出来，不能老是教书。以前如果在学校里教书，做设计的机会很少。那时候第一个念头就是去设计院，当时上海华东建筑设计院愿意接受，但后来因为上海户口政策的限制，不能同时携家属落户，最终没有去成。

正在此时国家恢复招收博士研究生了，当时童寯先生是南工首批国家批准的两位德高望重的导师之一。童老的学问确实很厉害，他的中文修养、古文修养、英文修养、艺术修养、绘画水平、建筑设计、学术论文什么都行，真的是一个非常全面的文艺复兴式的人物。他写的英文，南京大学外文系主任看了都佩服，他的水彩画展出以后，画家都佩服。所以他对学生的要求也很高。

华　您在读研究生或本科生的时候跟他接触过吗?

项　那时候接触不多，不敢接触。感觉不能望其项背，差距很大。结果博士研究生考上了，我也觉得很怕，为什么? 因为我们这一代人的小学中学的基础教育都是很浅薄的，他们这一代，包括圣约翰大学毕业的这一批人都比较厉害，所以我们觉得差距很大。在谈话时，他说到这个你也不懂，那个也不知道，我们就很无地自容。

华　童老要你们看书吗?

项　要，当初其实我们英文也不太好，看英文书也看得很累。童老要求说你一个礼拜来见两次，每次下午到我家里来谈一谈，你们有什么问题就跟我讲，"我是一个钟，不敲不响的"。所以，去之前你首先要看很多书才能谈，一个礼拜见两次怎么受得了! 后来我跟他的硕士研究生方拥说还是一个礼拜一次吧! 就是一个礼拜一次压力都很大。

华　书是他给你们指定的吗? 还是自己看?

项　自己看，他无所谓，非专业的都可以。你看了什么书，他就跟你谈什么，进行人文、文化、社会多方面教育。谈一个下午。

华　童老应该是非常希望能够把您带下去的。他也没有别的博士生吧?

项　没有。以前我们系里一个叫晏隆余的老师是指派给童老的助手。

华 我看到童老给您列出的研究题目，觉得很有意思，他好像特别希望您做上海近代史研究，他跟您交流过吗？

项 他跟我谈过，但没有说一定就做这个。因为当初“文化大革命”刚刚结束，我觉得一个人跑到上海各种部门去找资料，可能很困难，这个工作还要靠一个团体申请一个科研项目去做。作为一个博士研究生，一个人面对浩瀚的资料，凭一己之力，在当时的政治氛围下是很难搞的。

华 《长夜的独行者》中提到童老给您的几个方向，当时觉得哪个可以做下去？

项 印象中童老从来没跟我讲过有几个方向，大概是童文回忆的，我不太清楚。我跟童老接触时间只有一年，他后来就一直生病。基本上是他在病床上的时候，让我们把《古文观止》翻成英文，并给他解释一下。

华 为什么童老一定要选《古文观止》翻成英文，我觉得很好奇。为什么不是一个跟专业相关的书？

项 我觉得他这个想法从现在来看还是很前卫的，他觉得不能简单地把建筑看成一类工程技术，而是要当成人文的、社会的综合要素的表现，所以要成为一个好建筑师，文化修养这个底子一定要打好。我们今天也能很明显地看出来，做设计要有深度，要不落俗套，总归要有自己的思考的。所以建筑师的人文修养是非常重要的，培养一个真正好的建筑师其实不是培训一般的技术人才，而是整个人生的修养。童老主要选择中国古典文学，从最基础入门的《古文观止》开始。反正他的想法是要补课，尤其对于我们这批学生，1949年以后一直到“文化大革命”这一段，教育欠缺得很厉害，所以他等于说，你到了这个层次了，以后要读博士了，应该把这些东西补一补。

华 《古文观止》主要翻译哪些内容呢？

项 好像有几篇文章，其中一篇是《陋室铭》。

> 博士论文答辩，1985 年

华　这些书以前有翻译过吗？你们当时有英文参考资料吗？

项　好像我们没看到。我们自己翻给童老听了以后，如果还可以，他就点点头。如果不行就说这一句不应该这样子。压力真的是很大。童老的《东南园墅》是用英文撰就的。

华　是的，童老他自己肯定也很有感触。要把文言文翻成英文，就有一个跨文化的思想转换，原文的问题是中国的，英文会有更多现代的想法，这样就能把古今中外的思想联系起来。

## 在美国事务所工作感受很大冲击

华　项老师博士毕业，1985年就到同济大学来任教。所以您的经历非常具有典型性，虽然是高校教师，但是您并没有像大多数国内体制内的建筑教授那样，在设计院实践，走产学研结合的道路，而是很早成立了自己的个人事务所，充分跟市场接触。这是基于个人的选择，但是实际上个人的选择常常反映了某些时代的特征。比如就像您一样，1980年代有不少中国建筑师也选择了出国，我想这些出国经历跟您后面选择的市场化的实践模式有很大关系吧？

项　1989年9月，我从同济大学到美国做访问学者。一开始那边的学校还没开学，我住在旧金山一位朋友家，就想去试试能不能到美国事务所去干几天。那时美国经济不好，打电话过去都说“我们不需要人”，后来有一家在旧金山市区的公司，让我把作品拿给他们看，看了以后他们说“你的作品都太大了，应该到SOM这种大事务所去”，我说“大事务所都在裁员”，他们就让我先做做看。一开始做了旧金山市金融区一栋大楼底层的室内外改造，画了张草图，老板感觉很满意，就把我留下来了。

后来有一次我跟老板一起去见客户。当时他一边开车一边打手机，一面跟客户谈项目，一面又给公司里的人布置任务。当时我感觉美国的老板比国内设计院的院长难做多了，要自己开车，又要接项目，要到外面去跟客户谈项目，还要安排公司里面的人员。当时感觉在美国公司工作很辛苦，但又是对个人能力的一个挑战。今天国内外建筑师的状态差不多了，但在1989年，国内外差异是很大的，因此当初对我来说冲击很大。

1980年代国内还不允许有私人事务所，设计院的项目都是由国家计委分配的，不需要做市场，设计产值与工资也不挂钩。每个单位要盖房都需要由国家计委统一计划和立项，再由建委指定设计单位。设计单位不需要讨好建设单位，甚至设计单位因为要控制计划而成为老大。如果希望设计单位动脑筋超过划定的指标，建设单位还需要反过来讨好设计单位。比如因为我以前做过一个马鞍山富园贸易市场，做出了一点名气。昆山也想做一个类似的项目，昆山市的市长还拎了两条鱼，到我家来请我帮他们做设计，当时我自己感觉有些得意，感觉很受尊重。

在这家专门做室内更新的美国公司工作时，我也体会到了市场的残酷。有一次跟老板到一座要改造的高层去勘查现场时，我跟他一起到了二十几层以后，放眼四周一望，写字楼上全是空的，可见当时美国经济的萧条。后来这家公司专门招聘两个市场销售员，这些人都有绅士、淑女风度，精明能干，整天就是打电话给潜在的客户，约他们出来聊天吃饭，找生意。他们的工资很高。我们当时很不服气，想“我们干得这么辛苦，工资却没有他们两个做市场的高”。但是没想到过了一个月不到，他们两个就因为没有成绩被裁掉了。

华 当设计师就相对稳定一点吧？

项 设计师也很不稳定。因为市场化，老板是不养一个闲人的，一旦任务接不到了，公司的人就过剩了。今天去上班，老板让你到资料室把东西整理整理，差不多就是要炒鱿鱼了。当时有很多移民，比如俄罗斯或者欧洲其他地方来的，在自己原来国家也是水平蛮高的，作为技术移民移到美国去的。工作很卖力，起早摸黑，因为要拿绿卡和维持家里生活，工作不敢懈怠。我当时在美国孤身一人，没事情会在公司里待到很晚，另外一位40多岁的来自俄罗斯的女设计师也待到很晚，老板过来就说“I am moved that you devoted yourselves so much for the firm”（我很感动你们为公司做了这么多），尽管说得很好听，结果没过两天还是把那个女的裁掉了，事先没有一点预兆。

华 都很现实，很残酷。

项 后来就体会到你在公司里再卖力也没用，有活他就让你留下来，没活他就让你走，你永远缺乏安全感。我们国内出去的人开始不大能理解。
不过这种体制下，你个人要是努力的话，你就不怕，不是老板炒你，而是你炒老板，我在那里做了一年左右，后来就跳槽了。

华 项老师您当时为什么要出国？

项 在中国内地第一个获得建筑学博士，到同济大学当老师以后，我自己有一个愿望，希望在45岁之前能出国去看看，了解西方的动态。我觉得没出国就没资格当老师。将来的学生对老师的要求也会越来越高，当个合格的老师，要自己先去开阔眼界，首先就要有海外的经验。

华 您觉得教书比单纯做设计要求更高对吗？我们的第一代建筑师和老师，像杨廷宝、童寯先生他们都是留过洋的。后面两代就没有机会出国了。这些对您想出国有影响吗？

项 有很大的影响。因为我们当时到南京工学院以后，仰望的就是像杨老、童老这些前辈。他们学贯中西，知识渊博，而且积累了深厚的海外留学和工作的经历，具有国际化的视野和实践经验，这样的人才才能成为让学生受益的、合格的教师，自然是我们效仿的榜样。

## 香港的经历

华　在美国实践之后，回上海之前您到了香港，那是怎样的机缘？香港的建筑师是一种职业化的、服务性的状态，这段经历对您产生了怎样的影响？

项　当时香港的这家公司想请一个有海外工作经历，并具内地背景的建筑师。

华　为什么要请个内地的？

项　因为老板自己也是内地的，其实就是发展商请我去，本来叫我去当他的设计总监，我说我也不会管理，你要叫我来的话，还不如成立一个建筑师事务所，我们还是做设计，那是1993年到1999年。同时我也做公司的设计总监，也就是业主代表，有些项目发单请人做设计，我们负责监管。多数开发项目是自己事务所设计的，比如南京的江苏电信大楼，等等。

我可能也蛮矛盾的。我内心是想做一个比较有创新意识的建筑师，最好做老师，再做点实际项目。但是要生存，要跟上社会步伐，免不了要卷入职业生涯的漩涡中。自从离开上海以后，到了美国，到了香港，始终处于这种环境当中。最初是面临着生存的压力，要努力保住职位，不要被老板解雇掉。等到你自己开公司以后，你的事务所要能生存下去，就要保住你的“衣食父母”，不要被业主开除掉。所以我始终脱离不开职业圈的这个亘古命题。

在香港做的项目大部分是来自内地的，并且在美国和香港始终是操着非母语，按照当地的规范技术、行事方式工作。所以在当地人看来你都是缺点，语言不如我，对本地的情况了解不如我。在国外老是要去克服自己的缺点，尽管已经耗费了很大的精力，做了很大的努力。但是如果回国了，我懂英语，懂广东话，懂国外的一些操作方法，工作能力就相对强了。

香港的建筑市场有一个比较明显的特点，地方上一些普通的项目都愿意找香港本地建筑师做，高档的尖端项目都是找全世界建筑师做。内地建筑师在香港的市场很小，只能做本土或者内地的项目。所以觉得还不如回上海算了。

华　您接触到的项目哪种类型多一些？

项　各种类型的，什么都做。室内设计也做。

华　那时候要自己去跑市场吗？

项　那时候市场也是自己跑的。当时我们主要做高层建筑，后来觉得还是做内地项目好，香港的市场早就被占满了，你要参与不容易。

香港当初不承认内地学历，所以很多移民到香港去的内地建筑师，考不了香港的注册建筑师，而且因为不会讲广东话，只能在事务所里面做一般的工作。也有一些内地学生到英国留学了，但回到香港还是不行。香港的建筑师执照很难考。一帮子移民到香港去的内地人，有的人只能画画渲染图，有的人就做老板的打工仔。后来这群人聚在一起，搞了一个“香港中国建筑师协会”。这样香港建筑师学会的会长就着急了，找了中国建筑师学会来投诉，中国建筑师学会也派了人到香港去调查。

华　当时有多少人参加这个“香港中国建筑师协会”？

项　大概有四五十人吧。多数比我年轻，其实也是搞来解闷的。香港有一个规定，你如果不是考上香港建筑师的执照，在职场上就不能把自己称作建筑师，否则可以告你。当然，协会名称是另外一回事。

华　那您在香港工作的时候，有没有获得注册建筑师资格？

项　我有美国注册建筑师资质，在美国考的。

华　美国倒是承认中国学历的？

项　我去的时候也不承认。照道理我念到博士了，应该完全可以了。考美国建筑师有一个入门资格，要有5年美国承认的学历，再加上2年的工作经验。但是如果你没有5年的美国学历的话，则在美国学习和工作加起来要5年。我大概在国内到硕士研究生的学历，他承认了两年，加上我在那里工作的经验，加起来5年到了以后才给我考的。我九门考试全部考过了，考了几年，还有口试。

口试的时候，前面很多美国人都没考过，我想我也肯定考不过。三个美国考官问我的时候，我感觉英文表达得不好，回答得一塌糊涂，结果倒是考过了，可能这几个老头同情我。他们问很多规范，一个是设计规范，一个是加州的环境保护条例。这些知识我都恶补过的。

考完了就可以在美国开事务所，但是我不敢开，觉得拿不到活。反正中国建筑师在国外的处境还是不太理想。因为在加州的时候，我也参加了当地的中国建筑师活动，当时没想到它其实是个协会活动，就是礼拜六烧烤一下，然后就聊聊天，跟在内地的建筑师活动完全不一样。内地建筑师活动都要讲自己做了什么作品，有什么理念，美国那个中国建筑师协会活动都讲自己是怎么打工的。

其实美国建筑师自我感觉也还可以。因为在美国当一个建筑师还是不容易的。我在美国的时候，曾经问过事务所的老板，我说美国已经有这么多建筑师，也没这么多项目了，每年建筑学院还培养那么多人出来，怎么办？但是老板还挺为自己是个建筑师而骄傲的，说因此只有少数有能力的人才能当建筑师，在社会上也是被人尊重的。但在香港要差一点，香港建筑师即使是有执照的，也没有那么高的地位。

香港就是一个商业社会。给人感觉就是，拿项目第一，能赚到钱就行。

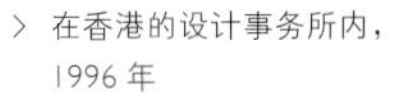
> 在香港的设计事务所内，1996年

## 成立上海秉仁建筑师事务所

华　您是什么时候注册自己的建筑师事务所的?

项　是准备从香港回上海的时候。

华　我注意到您的公司中英文并不对应。您一开始注册外资事务所的时候是同时有中英文的吗?

项　对,一开始叫都市建筑设计咨询有限公司,当时是作为留学生回来要注册公司,有优惠,注册资金很低,也不要求你一定要租下来办公室以后才能开公司。当时一开始选了"都市"这个名字,后来说不行;又搞个"现代建筑",又说重名了,后来我说就叫我的名字总可以了吧。注册外资公司也是用中文的,英文随便取什么都可以的。当时没有用英文,后来他们喜欢用英文就用了。

华　所以您最初并没有想要用个人的名字,后面用了也是机缘巧合。现在秉仁事务所是有资质的对吗?

项　对。一开始的时候国内也不批这种建筑设计专项的事务所。综合事务所需要相应的建筑师和工程师数量。我们光是建筑师为主的,只能批咨询资格,只能做方案。后来国内开始批准建筑师个人事务所了,我当时已经有了中国建筑师的注册建筑师资格。

华　您是在1995年直接给的,不是考试的对吗?可是您还没有回来。

项　我当时人在香港。一开始我也不在乎这个资质,后来深圳有一家民营事务所想让我做总建筑师,大概1996年或1997年,他们设计院帮我申报了国内的注册建筑师资格。

华　所以您1999年回来的时候,既有美国注册建筑师资质,也有中国注册建筑师资质。

项　后来国内开始允许批准单项建筑师事务所了,我聘用了两个校友做顾问建筑师,大概2004年批准了建筑单项甲级资质。

华　为什么您要去申请这个呢?因为有资质的事务所需要审核,是有点麻烦的。您的身份我觉得非常有意思。大部分像您这样资历的老师,或者是在学院,或者是大型设计院里,并不是真正的个人执业状态,因为中国建筑师总体来说市场化并不很完善,国营系统和民营事务所有着各自不同的资源,评奖和认同的系统都不一样,虽然最近一些年也在不断融合。您1999年回到上海,成立了一个那个阶段还不是特别多的个人事务所,当时是怎么考虑的?有没有考虑过进一个设计院做个总建筑师之类的?

项　这个跟对建筑创作的追求有关。因为在国际上,绝大多数事务所都是个人的专业事务所,比如建筑师事务所、结构工程师事务所、设备工程师事务所。我们国家在1949年以前也采用这种国际通用的建筑设计制度。建筑师、结构工程师、设备工程师分别开自己的事务所,有大的工程,三类事务所合起来完成项目,事务所和事务所之间有业务关系和专业上的关系,其中建筑师的主宰权比较大。

我们老一辈的建筑师都是从西方回来的，用的就是这种建筑师事务所的形式。后来我们因为计划经济采用了综合设计院的体制，至少从建筑师的角度觉得受到相当大的限制，不能主导整个设计。建筑师提的想法比较难做一点或者是比较费钱一点，同一设计院其他工程师会反对，建筑师就没有办法了。但是按照国际通行的市场化体制的话，做项目先找到建筑师，再找结构工程师和设备工程师来配合。如果建筑师要做，其他工程师说不行，我做不了，那你就不要做了。完全市场化的状态促进了建筑师可以提出很多新的要求，其他人要全力以赴配合他，这样工程师的专业技能也可以提升。我个人到美国和香港工作都是采用这种事务所的方式，因此我考虑让自己的事务所能与国际接轨。

华　建筑师事务所的全面市场化其实也需要其他工程师的市场化做支撑的。但是国内工程领域的市场化还很少，市场竞争还没有真正形成。

项　对，我们国内还没有真正市场化。美国的SOM也是有设备工程师的，不同的楼层里面有不同的专业，但实际上他们都是相对独立的。比如说建筑设计部门如果有一个项目，可以把这个项目后续的结构设备设计交给自己公司的结构、设备工程师来配合，也可以选择其他更合适的结构、设备工程师。

又比如说建筑师设计一个特殊类型的建筑，结构可能是另外一家更强的公司。作为项目的开发商或者是业主也可以选择，他的签约是分开的。有了专业和专业之间的相互制约和专业的竞争，才会出现设计行业的进步。这一点很重要。在香港的时候，巴马丹拿这家老牌事务所，有500多人，里面分建筑设计公司、结构设计公司和设备设计公司，如果自己公司的结构设备做不到建筑师的要求也可以被淘汰，一个公司内部也可以这样。

华　所以是有竞争的，建筑师有选择的余地。但我们设计院系统就做不到。

项　现在受到一个设计院的管理限制比较难做到。但也不是做不到，设备工作吃不饱、项目不够的话，做设备的工程师会到外面找活干，因为已经是单独计算产值了，现在这一点放开了，以前是不可以的。比如华东院以前和我们做项目合作的设备所也会来找我们，希望我们有大型项目可以跟他们配合，建立一个长期的合作关系。大型设计院也在变化。

华　您在做不同项目的时候，会选择不同的配合设计院吗?

项　对，两方面考虑：一方面是看技术上是不是最适合做这个项目的；第二个是价格上面。

当时我们做合肥大剧院投标中了标，业主是政府，对协议就有要求。我当时在同济，不好意思把项目给别家做，就跟同济设计院合作，但同济设计院跟甲方谈下来一个费用，之后就按照设计院的内部分成制度分配建筑这一块，我们的占比就很少。我们前期这么努力投标他也不管，就是按照他们内部这个比例来分配。

## 事务所的运营模式

华　这对国内独立事务所有很大的制约，因为其他技术没有完全市场化，所以选择的余地相对比较小。设计院自己的项目也很多，没有那么多精力分到市场里。另外，项老师这边的项目，酒店、商业别墅、养老、商业地产是主要类型。

项　这就要谈到我们的设计师事务所的定位了。建筑设计既属于艺术创作，也是商业行为，市场当然是非常重要的一个方面。但是建筑师个人也有很多自己的追求和理想，完全把建筑设计看作是商业行为的话，就没有意思了，这个不是我的理想。

一直以来，建筑学是自己的热爱，我觉得即使是工作也要有兴趣，如果工作虽然赚钱但是很乏味也没有什么意思了。所以回到学校以后，我希望能通过教学和个人工作室来参与社会活动。原来我们的公司也是工作室的概念，一开始没有这么大。我自己的想法是既然是工作室，一定要有创作性，也想争取一些好的项目。我比较喜欢避开过度经济性的考虑，可以做一些建筑性的事情，也不能只停留在纸面上，能力要在建筑过程当中得到提升。

刚开始工作室固定员工只有两三个人，还有几个研究生。研究生希望参与实践，得到一些不仅仅是纸面的教育。学生选老师的时候也会希望老师有一些个人的实践。这个情况全世界都一样的，建筑学的老师不太可能一个人做设计，也不太可能纯做理论，因为你是建筑设计，总归还是要有实践，学生也不希望老师整天是空谈的。

华　项老师，现在我们事务所里是以前期设计为主，施工图主要是跟其他设计院配合的是吗？

项　是做到初步设计。

滕　虽然说是做到初步设计，但这个阶段其实已经有其他工种的配合了，我们统筹把各工种提出来的条件放到图纸上。我们把外立面控制得很细，基本上是可以指导幕墙公司进行设计的。如果是一个窗墙体系的建筑，基本上我们的图纸可以直接拿去用作施工图了。

华　除了外立面，室内设计会控制吗？

滕　室内做到初步设计的阶段。但我们会做一些专门的室内项目，多数是文化建筑，因为想要从外到内都控制，对空间里面的一些材料质感有自己的想法。这也是项老师本人的兴趣所在。但其他类型的就不太会做深化。

项　其实按照国外的做法，施工图也是交给施工单位深化的。设计人员画出来的施工图，跟施工现场的技术发展程度已经不配套了，学了半天画出来之后，到了工地也只能重画，何必浪费这个时间？比如说我们做大剧院的设计，我不可能把大剧院设计的所有技术问题都花很多时间去研究清楚，具体的一个面光、一个耳光的角度，观众厅的视线、声学等问题，没有精力一一去研究啊。所以没有必要去做剧院专家，什么都要懂。建筑师还是要抓住关键问题，把它的空间和设计特色研究好。

华　项老师，您当时在美国时，那些建筑师事务所会做到施工图吗？还是直接交给施工公司去翻样？

项　中小型事务所都不会做施工图。

华　在香港的贝斯也是这样?

项　也是以前期为主。包括前期去跟业主沟通，落地到方案，再到扩初阶段，后期主要是空间效果的控制为主，基本上不太会涉及具体的施工图。
我做施工图就是在我大学毕业以后，相当于是在设计院。功能比较简单的项目有做到施工图，但后面大型的项目基本上都是以前期为主。以扩初为主，等于我做到了一个建筑学的控制、材料空间的控制，比较符合感兴趣的建筑学的部分。

华　我看到项老师您的工作室里，下面的主要骨干是以同济毕业生为主，以前我们以为只有高校设计院或教师工作室里师生关系会比较多，但之前德国GMP事务所中国区代表吴蔚先生做了一个讲座，讲到GMP二十几个合伙人中，除了三人以外，全是GMP三个创始合伙人的学生。我不知道您这边是不是也有学生进入事务所，形成其他事务所无法比拟的源于师生关系的思想上和感情上的延续性?

项　应该说是根据人才的选拔。

华　您选研究生的时候会有什么特殊标准吗? 对于学生来讲也是双向选择的过程，选择哪个老师实际上是选择了自己未来的方向。

项　应该是这样的。我是通过学生阶段的工作了解他们的能力、人品。东南大学、同济大学的学生可能会比较知道我，找到我这边来。就我个人来说，也认为这几个老学校的学生比较好。

华　您作为老师，跟事务所在经营上有什么关系吗?

项　理想当中，我自己是一个建筑学的教授，也是一个建筑师，同时有自己的工作室，绝对不会去开设一个大的设计院。但是开事务所就会面临市场问题，遇到好的项目需要人却没有人做，就很着急，就想扩大自己的事务所。扩大了事务所就得维持，不能把人裁掉，只能继续扩展项目……不过还是希望在创作方面有可以发挥的项目，不单纯是完成生产任务。

华　具体操作上，是不是每个项目有负责人，您需要所有的都知道并把关吗? 一开始的时候是您自己创作为主的吧?

项　以前我每个项目都要管。我在香港的时候听到一些关于刘荣广和伍振民事务所，他们在香港做得比较大，每年有100多个项目，香港中环大厦都是他们做的。刘荣广曾经跟人家说，100个业主来跟我讲他的项目，每个项目的业主和情况我都要了解，这就是商业建筑师需要做到的。所以我觉得既然是自己的事务所，没有能力做100个项目，做50个项目，50项目也行，但要对这些项目都知道。

滕　补充一个信息，项老师原来在公司的时候，更多是扁平化的管理，不像商业事务所一个部门带领几个团队，架构是金字塔型的，各自团队里是层级化的管理。项老师管理项目的时候会每个人都管，下面的人他也会顾及到，比如你这个画得怎么样他研究一下，那个人设计得怎么样了他研究一下，倾注了很多心力。也是因为这样的原因，公司不可能无限膨胀，一直是20个人到40个人，对他来说失控的感觉是他不想要的，他想每个项目、每个设计都掌控住。

项　有点这样。或者这样讲，大学老师还是需要让学生有点崇拜，这样才能得到好的教学效果。什么是好的教师，有很多考核指标，考核老师上课不迟到？下课在教室里很勤劳？教学工作量很高？很多时候学生一毕业以后把老师忘了，但是你看冯纪忠老师，走到教室一站，一大批学生就围上去了，他讲的每句话别人都记得住，这才是好的老师。要成为好老师当然个人要有魅力。

华　用人格和学识影响别人。

项　冯纪忠老师除了理论以外，他有作品，又能画，文学底子好，英文又好……学校想要真正达到教学效果就需要有一群好的老师，制度和经营没那么重要，好老师才是一个学校的财富。我当时回学校也希望成为能被学生崇拜的偶像。我能力不够没有做到，但是目标是这个，所以工作也是朝这个方向努力。

作为一个建筑系的老师，肯定自己要有作品，让学生看到这个老师可以做出东西来，进一步也会影响着学生。本来开事务所只是把它作为自己的工具，等到真的有了DDB以后，才觉得很难。在中国改革开放的经济大形势下，很难成为一个完全不营利没有经济支撑的事务所，如果只是一个老师带几个学生，靠老师的经费做事情也做不起来。既然希望可以稳定做点项目，那还是要有能够维持自己的费用，包括人力资源也需要费用。这样的一种事务所在大环境下，不得不参与市场竞争，纯粹的工作室是很难做的。比如有些项目，业主说要考察一下你事务所的规模怎么样，他会看有没有一些人天天坐在办公室里上班，打电话过来能不能找到人，最后还是要有公司的样子。完全是一个工作室的形式肯定是有局限的。

国内有一段时间流行实验建筑师的概念，他们大多是做一些比较小型的、商业价值不太高的建筑，但是我也不认为这些建筑师是想永远停留在研究室的阶段，可能他们也是想通过实验建筑师的业绩来打开知名度，开拓自己的专业领域。建筑师不是一个完全艺术性的专业，而是艺术和技术相结合的专业，所以能不

> 在上海的设计事务所内，2007年

能把握住一个大型的开发项目，同时也能够把你的建筑概念表现出来，对于建筑师来说是很大的考验。实验是可以的，我觉得不能单纯迷恋做小型的东西。

华　这涉及您对建筑师身份的定位问题。有的人做小的实验性设计，在意的是体现自己的概念，不会把服务社会作为很重要的目标。而您对建筑师的定位是建筑师本身是需要为社会服务和作贡献的，同时把自己的学术理想融合在对社会的贡献中。您在网站上有一段话讲到建筑师和业主的关系，说建筑师是业主的参谋，建筑师不能以业主为唯一导向；另外您觉得建筑师是去解决问题的，而不是为业主制造新的问题的。我认为这里面就有一个自我定位的问题，不希望为了表达自己个人的东西就盖过业主本身的需求，是这样的想法吗？

项　我觉得建筑不是单纯的艺术，建筑一定要跟社会联系在一起，所以建筑师如果要发挥更大的作用，应该是把你的先进理念通过实际工程体现出来。你在做一个大型项目的时候，整个过程会受到各种各样的挑战，政府的规范，工程的、技术的限制，还有各种各样的专业要求都要解决。在解决这些复杂的综合性问题的同时，建筑师还是可以一步一步地把建筑最初的理念坚持下来并实现。这需要通过不懈的努力和毅力来完成，同时也要有一定的专业技巧。

华　大型项目相对来讲功能的需求或者其他技术方面的要求就会很多，不像小型项目那样可以很理想化地贯彻建筑师自己的概念。您选择了大型项目，实际上已经选择了一个不是完全以建筑师个人的概念为主，而是一个以更综合的考虑为主的状态。

项　建筑师长期积累的经验很重要。比如说在纽约贝聿铭事务所里，贝聿铭跟我谈话。我问，你现在中国的项目能不能去做？他说，兴趣是有的，但是有两点困难：一是坐飞机，一年到中国能走三趟最多了，做个项目绝对不可能跑三趟就可以解决问题的，长途飞行受不了；二是缺乏助手，助手能够帮你处理所有的问题。从这点上可以理解，一个建筑真正要做好的话，除了概念之外的整个过程，不是助手就是他自己必须具备驾驭这个项目、处理各种各样综合复杂的专业问题和管理问题的能力。所以我觉得，凡是这种大师都有过这类经历，只需要指点江山了是因为后面的人可以独立做了，有这种管理综合项目的能力了。建筑不是单纯的纯艺术的问题。

华　对，实施层面有很多的实际工作，跟各方面的协调，等等。这对项目建成来说是非常重要的，如果做纸上建筑是相对容易一点。

项　东南大学的王文卿讲过一句话，他说“建筑老师”与“建筑师”，差一个字差了很多。建筑教师可以不需要考虑实践，不需要考虑做生意；但是建筑师要会实践，还要考虑自己的市场定位和经济状况。
讲回我们的DDB，市场定位是很难的一个事情，因为现在的市场是不太区分的，什么都可以做。这当然有一个好处是我们可以接触很多项目，有很多机会，但是有时候也不是我们公司的力量可以做到的，不过我们现在基本上是都做的。

华　也跟市场有关系，或者是前期已经这样，您再细分的话就要取舍。

项　我一直希望我们的项目能够做得更好，不只是满足业主的需要，也希望业主可以抬头仰望你，而不是低头俯视你。所以我说一定要做精品，为了做好的项目一定不要怕做投标，现在被别人请去投标是对你的地位的认可；不要在乎是不是第一名，全世界那么有名的建筑师像库哈斯这些人都参加投标的，不中标的也很多，这么有名的照样也有不中标的。我们国家现在有一个不太好的情况，就是都要讲头衔。

华　因为技术上很难划分，所以只看头衔；技术上的差异官方觉得不大，从质量上也都不错，不如找个有名的，对后面的宣传会有利，类似于一个市场行为。
设计系统现在已经分好了，比如说民营的公司主要是做房地产市场的，国营的系统走官方路线，无论是项目来源还是申报奖项或是资历的积累，都是系统的资源。市场化的得不到主流的资源，某种程度上两者是分开的。

项　对，这个也没有办法了。

## 设计管理制度和评奖体系的困惑

项　实际上现在国内有点混乱。国际上的做法有两种：一种是大部分项目由建筑师设计事务所总包。香港的AP（Authorized Person）就是项目的主要负责人，如果没有考上AP，你就不能在图纸上签字。在香港你考上建筑师没有用，还必须得考一个AP，才能够总包项目。这种人的权力很大，包括工程进度款需要AP签字以后业主才给钱。
我们现在做一些项目，事务所单独和发展商签约，签的是前期设计这一块。发展商找设计院，设计院负责画施工图，除了建筑专业之外还有结构、设备这些专业，这些专业的收费是另外一个标准，他们另外谈的，也不需要经过我们同意。按道理是应该要经过我们同意才可以签字的。这样一方面我们建筑师可以更好地对项目负责，另一方面也能够使配合单位、配合工种服从建筑师的总的意图。

华　但我们没有这个权利。

项　过去建筑师权利很大，特别是在香港和1949年以前的中国，承包商要拿到工程款一定要建筑师签字。制度规定了，合同上也是这样写的。建筑师要负整个项目的责任。

华　建筑师责任制。您说香港的情况，是指签约的时候可以分开合同，有建筑的合同、有其他工种的合同、有施工企业的合同，不一定都归口在建筑师这边签订，但在管理上是必须要通过建筑师签字他才能拿到钱，对吗？

项　对。

华　现在的施工部分有些修改是让建筑师签字才能从业主那里拿钱的，其他的可能没有这么多的限制。建筑师就很难对整个项目起到主导作用。

项　建筑师的权利和责任是同时存在的，赋予建筑师这个权利，使得你的设计意图得以高质量执行，但是建筑师也有责任，如果出了问题要承担相应的责任。

华　这和收费标准应该也有关系吧。现在我们国内也在推行建筑师负责制，但是建筑师提出来，我收的费用那么少，为什么还给你负责？某种程度上，国内的建筑师即使没有收相应的费用，其实也是在负责的。建筑师作为建筑行业的总控身份，也不可能完全放弃。

项　但是合同的规定比较模糊，没有制约也没有权益。

滕　我觉得还有甲方管理和意识的问题，同时也受限于公司其他专业技术能力的限制。成熟的甲方管理团队会要求建筑师进行建成效果总体把控，需要到现场落地去控制，甚至还要有正式的合同备书。而有一些甲方可能就没有这个意识，甚至觉得只要你能完成一张漂亮的方案图就可以了。现场出现问题了，不是通知原创设计单位去解决问题，而是图方便找施工图单位或者现场施工队草率应付。就责任方面来说，建筑师审查施工图，但无法对其他专业的所有技术问题去承担责任。如果这种责任制需要白纸黑字落尽合同里，可能我们自己就入坑了。

项　美国建筑师协会的法规里就是规定建筑师仅对完成效果这一块负责，其他的不负责。

华　结算、结构、设备工种各有规定，建筑师对其他专业本身的技术问题肯定不能负责。我们最终的工程质量，比如说结构设备影响到建筑的空间或者使用功能的时候，建筑师肯定要负责。我觉得现在最大的难点，一是因为没有充分市场化，建筑师没有办法自由选择结构和设备；另一个是建筑师在权利上对施工公司没有制约权，因为经济上不挂钩，没有统筹，导致没有办法控制，最终造成建成成果质量不好或者效率低下。
我还想问一下项老师，您觉得单项资质的建筑师事务所的模式，在国内开展业务、评奖时，跟综合设计院相比，渠道有什么区别吗？

项　现在从制度的层面看还是不尽如人意。比如很多项目的招投标一定要有设计院的综合资质，一投标发现你不会做施工图，就把你排除在外了。但现在也不能一概而论，有的开发商财大气粗，当地政府比较配合，技术部门说请外国的事务所，就可以进入了。国内现在慢慢有一些大牌的、比较著名的事务所也被接受了。
再谈谈评奖方面，国内的奖项背后的体系也不大相同，有的是建设部的体系，有的是民间的体系，比如说建筑学会的体系。我是单项甲级资质，上海的可以参与。据我了解，浙江、江苏、安徽这些地方基本上还是需要综合设计院才能报，评奖的通知只发布给建设部下属的设计院。有些项目我们知道是外资公司做的方案，国内设计院做的施工图，奖项报出来却是中国设计院得奖，一点侵权的感觉都没有。我们也有遇到过，比如我们设计的一些大型公建项目，是合作的设计院在当地省市得奖的。

华　按理说应该算合作设计的一个奖，他们没有把你们放上去？

滕　设计院报奖的话，可能会自己单独报，或者是把我们的参与人员写在他们后面。

华　我们现在体制里一些个人头衔，比如勘察设计大师，如果没有这些成果也无法申报。

项　对。

华　你们会系统地参加一些社会上的报奖吗？

滕　会的。现在社会上商业的奖项比较多。这样的奖项有商业运作在里面，控制奖项的可能是甲方或开发商，跟甲方的项目有很大的关系。其实是甲方在评选设计师，因为大部分评委可能是各个地方开发商的老总。

华　这个是另外一个系统，以商业为导向。

滕　对，算是一种自娱自乐。地产类的项目很多都是类型化的，成熟的地产开发更注重价值回报，更注重故事性，跟学界的奖不太一样。比如前几年大家都谈新中式，但是我做的和你做的有什么不一样，讲故事的能力可能就比较重要。

## 为每个项目做不同的创作

华　有一点我一直很好奇，东南大学的教授，比如齐康先生，在1980年代后期设计了很多文化建筑，有一定的纪念性。但项老师的作品，或者您关注的大部分是城市中相对现实和日常的类型，比如办公、商业等，您为什么比较少做文化类、纪念性的项目，这是有意选择的吗？

项　不是不想做，是没有机会做。每个人都有自己的方法，但其实我不敢说形成了自己的一套建筑思路。对我来说，总是想做与众不同的事情，是自己内心的一个愿望。建筑有很多是实用的考虑，在有限的资源里，每个人都会去寻求一个答案，十个建筑师就有十个答案。你怎样能跳出来，比其他九个建筑师有更加独特的想法？这是我比较追求的，虽然不一定每次都追求得到，但有时候在项目投标当中会想去体现不同于常人的应对问题的方式。自己要形成固有的一套设计方法、设计思想、设计逻辑，真的很难。因为遇到的项目也是各种各样，你很难拿一种观念、一种处理方法去解决所有问题。

华　这跟您作为职业建筑师，接触的项目类型比较多有关吧？针对不同的类型，您肯定不是先入为主的。

项　有些签名建筑师（Signature Architect），比如扎哈·哈迪德，随便什么类型，办公楼、住宅、剧院、美术馆，都是一样的，一看就是扎哈的东西。要做签名建筑师也很不容易，因为要有一套自己的造型语言，要跟人家不同。虽然他们拿同一种形式、同一种方法来解决所有的问题，但是他们这种形式也是多年苦心钻研的结果，所以不能小看他们。

华　您觉得签名建筑师是一种形式主义吗？

项　不知道能不能扣一个形式主义的帽子？因为它的形式确实能够用到不同的场合里，它还是有变化的，大的风格是有形式特征的。但有形式特征也不等于是形式主义，因为不是一定要把其他某种形式套在上面。比如说浦东的平安保险大楼就是形式主义，搞一个固定的、套用传统古典的形式，不是自己的建筑语言。我倒是蛮想成为签名建筑师的，但是我没有这个修养，没有这个积累。其实美国有很多大学建筑系的教师，搞个小事务所做研究，哪天搞出来了，被人承认了，他就出名了，就开始有很大的市场和机会了。譬如路易·康就是这样子，开始的时候也是默默无闻的，在小的studio（工作室）里研究，等他确实能够说服大量的人接受他的观念和形式，他就成功了，别人也比不上他。虽然我也蛮想成为这样的人，但是我没有这样的条件。既然我们不是签名建筑师，那就应该根据不同的项目，跟不同的业主打交道，处理不同的项目要求，建造在不同的环境上面，那就应该根据每一个项目进行创作。

## 建筑学教育、研究生培养等

华　谈到签名建筑师和职业建筑师，现在国内的建筑学教育中，职业化的教育比较少，成为明星建筑师感觉是最高理想，但实际上学完走上社会以后，如何成为一名职业建筑师反而更重要一些，这个情况您怎么看？

项　其实我们国家现在的建筑教育好像没有一个比较科学的分类。社会上需要很多职业建筑师，但是我们同济大学是不是该培养职业建筑师？在美国，比如说加州理工大学是着重培养职业建筑师，毕业出来的学生去考建筑师，一考就考上。设计单位也特别愿意要这些人，去了就能干。职业建筑师把美国的建筑设计的基本技能、方法、规范、技术等一套职业化的训练都掌握得很好，一出来就能工作得很好，也能满足社会大量的需要。但也有一些学校就是培养尖端的、精英化的人才，比如我以前去过亚利桑那建筑系，虽然不大也不太有名，但是他们教授说，我们就关心学生毕业后能不能进入美国最大的公司，如果有两到三个学生在社会上有名了，我们的教学就成功了。也就是说他们的目标就是培养高端的精英化人才。不过国内的情况有点搞不清楚，譬如在同济当老师的时候，一个班上三十几个同学上我的研究生设计课，以后可能最多1/3的人做设计，1/3的人到开发商那里去了，还有一些人做其他什么事情。不是所有人都对建筑设计很感兴趣的。而作为老师都不知道这些人才将来派什么用处，你怎么来设置教学大纲和教学计划？现在学校里输送出来的毕业生，到底跟外面的市场能不能对接？比如我培养了好多博士，不是都进入研究机构的，这就是学历泡沫，是浪费。

华　您现在培养的研究生，有多少是像您一样很专注实践的设计师？我不知道您在研究生培养时是不是比较希望关注设计前沿的问题？

项　学生培养这一块我是比较开放的。

华　怎么计划的，根据他们自己的兴趣吗？

项　我回到学校以后也开始考虑应该向哪些方面进一步开拓。原来我的博士论文研究的是城市设计的问题，

> 与学生合影，2019年

大概是1985年。后来我在同济大学开过一门课，就是讲解我的博士论文。现在规划系的研究生找在同济最早开设城市设计课的老师是谁，他们说是我，我说我也不敢当，因为我当时博士论文写这个，所以搞这个比较早。

但是现在我有点失望，因为城市设计好像变成建筑系老师的一个研究的大方向。光是建筑学的研究很难上升到理论高度，局限性很强，于是把建筑设计和社会、环境、历史结合起来，马上都能有比较多的研究性内容了，所以很多老师报科研方向的时候都报了城市设计，好像很先进，但是城市设计真正解决什么问题也不知道。我出国十年以后回国，回到建筑系，听到建筑系老师报的科研项目都是城市设计这方面的内容，心想是不是把城市设计作为一个科研方向变成敲门砖了，我觉得这样没有什么意思。

而且城市设计最终能够实现还是要靠好的建筑设计，有觉悟的建筑师很注意自己的房子在城市中的地位，与城市建筑的关系，所以我后来不愿意再把城市设计定位为自己的方向。我出国十年做了很多设计，去了很多地方，所以觉得应该更开放一点。如果考虑到时代发展的需要，数字化建筑应该是一个方向，因为是跟网络计算机、智能化时代联系在一起的。我于是倡导数字化建筑作为研究的方向。我看了一些资料，但是这个方向我做得很少，写过几篇论文，不像现在袁烽搞得很好。

对我的学生，我主张不要把他们作为老师科研的劳动力，而是要尊重学生自己的爱好。大学本来就是一个自由思想的地方，如果自己有一个很强大的成体系的科研计划，你需要有一帮人帮忙，可以把学生纳入到你的系统里；本来你还没有很完整的体系，我希望学生每个人去发挥自己的特长、自由追求目标。所以我的研究生的论文都是看他们自己的兴趣和爱好，跟我一起讨论，觉得这个题目有意思，适合你的需要，就可以做。在整个过程当中学会怎样做科研，怎样写论文。至于最终他的研究成果能不能跟他以后的职业联系在一起，这是一个大问号。

华　您在一开始的时候，是邀请研究生来参与您的设计实践的吗？

项　自愿的。想过来的人就过来，不想过来的就不过来。

滕　可能会有机会，项老师说这里有个项目你们有没有兴趣来，有兴趣就来，也有同学不来的，也没有关系的，不会对他/她有什么影响。

华　项老师，您1999年刚从香港回到上海，回到同济。我不知道您还记不记得，2000年您来参加我的硕士论文答辩，我做的是邬达克建筑的研究，您问了一个问题，邬达克是明星建筑师还是商业建筑师？您这个问题跟别人的都不太一样。那时你刚回上海成立了自己的事务所，我不知道您当时会不会也在思考自己事务所的定位问题？

项　这个问题考虑过。其实现在作为生活的乐趣来说，当然愿意成为明星建筑师。因为你不是整天扑在图板上，不是整天做一些很乏味、简单重复的商业经营的事情。我个人还是愿意搞艺术多一点，所以还是比较向往明星建筑师的这种状态。我年纪轻一点的时候，还有一点点可能会达到，每年都有机会。但现在我不是明星建筑师，现在明星太多了。我为什么问这个问题呢？因为我觉得邬达克肯定是一个商业建筑师，他不是一个明星建筑师，当初他那个时代，就是开一个事务所来接项目。

华　虽然邬达克设计做得很好，但是您会把他归类成一个商业建筑师，是因为他是服务型的，以满足业主为导向，没有自己明确的立场吗？

项　因为建筑学之所以可爱，就因为它跟一般的技术工作不一样，没有那么多重复的劳动，有创意的成分在里面。所谓做明星建筑师就是说你不断在坚持建筑学本身的发展上面多动脑筋，而不是说做生意怎么做得更大，为赚更多的钱去动脑筋。

华　所以您所指的明星建筑师其实有一个更大的诉求，要在学科上有更大的贡献。

项　商业上的成功其实很乏味。你看今天的城市里，盖的房子都是似曾相识的。一些大型设计公司因为产值和营利目标的驱使，生产出了很多规模宏大的商业和居住项目，公司的经营压力和运作机器基本主导了员工的工作和生活方式，因而难以在建筑设计方面有所创新。而另一些由建筑师主导的设计事务所，由于有自己的专业理想，能在一定程度上摆脱营利的纠缠，反倒能扩大自己的专业视野，跳出平庸的巢穴而做出一些有想法的设计来，同时个人也能享受工作的乐趣。这就启发我们该怎么对待人生，如果你过于刻板、过于拘谨，循规蹈矩，可能会获得一般意义上的成功，但如果你更加潇洒一点，不管社会常态如何，选择适合自己的生活道路，也许会更开心，我是比较喜欢后者。很多东西不一定正确，但是至少随自己的心去做事情。

华　比较个人，比较佛系一点的状态。

# 1

雨山湖公园小品建筑
Landscape Architecture of Yushanhu Park

马鞍山富园贸易市场
Fuyuan Market, Ma'anshan

昆山鹿苑市场
Luyuan Market, Kunshan

胡庆余堂药业旅游区规划
Hu Qing Yu Tang Pharmaceutical Tourist Area

## 早期城市建筑设计：现代主义与折衷主义

动荡的十年“文化大革命”之后，项秉仁终于如愿重回到绘图桌前，并开启了他的建筑师生涯。项秉仁在这一时期，无论是观念上还是方法上都在追随和实践现代主义的建筑思想，马鞍山雨山湖公园小品建筑（1976）就是其中的一个实例。

而到1980年前后，即项秉仁回校攻读研究生的这段时间，后现代主义建筑思潮在中国流行。这股思潮对项秉仁原本毫不怀疑并视之为金科玉律的现代主义建筑观念造成了不小的冲击；加之赖特建筑思想中原本就潜在的反国际主义意识，以及他的博士研究课题所具有的时代性，同时对这种冲击起到了推波助澜的作用，结果造成他这一时期的建筑设计呈现出折衷态度。这在富园贸易市场（1984）、昆山鹿苑市场（1986）、胡庆余堂药业旅游区规划（1987）等项目的设计中表现得尤为明显。这三个设计都是有关于城市片断的探索，涉及街道和广场等城市空间，传统建筑文化元素得以强化和裂变，带有后现代主义的表现性。

## Early Urban Architecture Design: Modernism and Eclecticism

After a nation-wide turbulence of ten years, Xiang Bingren was finally able to return to the drawing table as he wished and started his career as an architect. He followed modernist ideas both conceptually and methodologically. An example from this period is a small architecture in Landscape Architecture of Yushanhu Park, Ma'anshan (1976).

Around the year 1980, when Xiang was back in school to pursue a master's degree, post-modernism began to gain popularity in China's architectural world. This new trend had a great impact on the modernist concepts that Xiang had regarded as the golden rule, an impact that was encouraged by the hidden anti-internationalism in Wright's architectural philosophy and the awareness through Xiang's doctoral research. This resulted in a trait of compromise in Xiang's design, as evidently shown in projects such as Fuyuan Market (1984), Kunshan Luyuan market (1986), Hu Qing Yu Tang Pharmaceutical Tourism Area plan (1987). They were all about exploring city fragments, involving urban spaces such as streets and squares. Here, traditional architectural elements were enforced and then divided, displaying the representational feature of post-modernism.

建筑与自然的融合 A pavilion with its greenery surroundings

# Landscape Architecture of Yushanhu Park

1976 / 1976

# 雨山湖公园小品建筑

在“文化大革命”结束数年之后，历经南北辗转的项秉仁终于在安徽省马鞍山市建筑设计院开始了其职业建筑师的生涯。当时马鞍山这座新兴的钢铁城市正处于如火如荼的城市建设之中，项秉仁在同期还设计并建成了其他各类建筑作品，如马钢医院制剂车间、马鞍山市电视台、马鞍山市建筑设计院大楼等，但遗憾的是，由于时间久远，加之中国城市更新速度太快，仅有雨山湖公园小品建筑的资料还保留至今。

雨山湖是马鞍山市中心的一处自然风景，市政当局围绕着这一景观资源开发了雨山湖公园。公园内有大量绿化和广阔水面，小品建筑点缀其中，既可供游人驻足赏景，又与自然山水林木相得益彰。项秉仁在设计中运用了大量建筑和结构的现代主义处理手法，使建筑与环境相互融合。建筑临架于湖面之上，有一种舒展的漂浮感。架空的底层和回廊很好地将内外景观互通。独立支柱、内部空间中不时出现的“片墙”、暧昧的空间隔断都体现了现代主义的影响。悬挑于水面上那个无柱支撑的楼梯，不仅让整体建筑轻盈且连续，也能让人体会到“结构表现”的趣味。

我们不难看出，这一时期的项秉仁深受学生时代的建筑思潮影响。建筑设计作品在观念和方法上都追随和实践着现代主义的建筑思想，如形式追随功能、空间流动、表现结构等，而在技巧上则体现出他在南京工学院建筑系获得的来自“布扎”体系的基本功。

---

After a period of transiency in the years following the Cultural Revolution, Mr. Xiang finally found a place to launch his career as an architect: the Ma'anshan Architecture Design Institute, Anhui Province. The city was experiencing rapid development at that time, as the local steel industry contributed to unprecedented city construction. Due to this booming growth, Xiang Bingren participated in various architectural design projects that might not have been realized had the city not been in full economic swing. He participated in the pharmaceutical plant for Ma'anshan Steel Group's Hospital, the Ma'anshan Television Station's Headquarter, the office building of the Ma'anshan Architecture Design Institute, among others. Unfortunately, the design documents of all the aforementioned projects were lost to time and the rapid, often chaotic redevelopment of the Ma'anshan city center. Only documents from

廊道空间与湖上平台 View of the leisure platform

several landscaping projects have survived.

The municipal government of Ma'anshan City built Yushanhu Park around Yushanhu Lake, a natural body of water with beautiful scenery located adjacent downtown Ma'anshan. The park features ample greenery and open water, while providing locals with beautifully-landscaped views around a scenic lookout without interrupting the natural environment. Thus, Xiang Bingren extensively used modernists' design tactics to enhance the interaction and infiltration of architecture and nature. The landscape piece rises above the water elegantly, as if floating on air. The lifted ground floor successfully blurs the boundary between interior and exterior. Sporadic walls serving as ambiguous spatial barriers hint at the subtle. The most stunning part of the design is the staircase hanging freely above the water, emphasizing a consistent levity inherent in the design along with nuances of "structural expressionism".

It's easy to see how Xiang Bingren's designs, at that time, were under the influence of architectural trends prevailing during his schooldays. He practiced modernist prerogatives such as "form follows function", spatial mobility and structural expressionism both conceptually and methodologically. Xiang's designing techniques revealed his "Beaux Arts" training from the Nanjing Institute of Technology's Architecture Department.

室外悬挑楼梯 Open cantilever stair

回廊 Covered walkway

## 评论
## Review

雨山湖公园小品建筑是项秉仁老师的早期建筑实践。他充分利用了钢筋混凝土的结构特性，竖向采用了墙体、独立柱的结合，顺应空间和景观组织的需求，布置灵活自由，廊内空间流动而丰富。梁板的关系也十分微妙，梁只出现在外围，廊内的柱间并没有联系的梁，这使水平的结构板获得了一种轻盈感。建筑沿湖展开，平台及楼梯外挑，悬浮在湖面上，进一步强调飘浮感。在立面设计中，他采用了纤细紧密的混凝土格栅、简化的挂落等手法。其中，靠近地面的花坛被设计成为悬浮的姿态，也是极为有趣的结构性的处理方式。

雨山湖公园小品建筑呈现出许多早期现代主义建筑的特点，但同时体现了中国园林建筑的空间意趣。如果说中山陵音乐台是1930年代将中西方古典建筑相结合的园林建筑的经典，那么在1970年代，雨山湖公园小品建筑则是蕴含现代主义建筑理念的地域性探索的典范。虽然当时建设条件十分有限，但在这样的环境中，项老师的实践更多地着重于对建筑本体的探索。

郑泳
具集建筑主持建筑师
建筑学博士

The landscape piece of Yushanhu Park is an early architectural practice for Xiang Bingren. He utilized the characteristics of reinforced concrete structures to the fullest. The combination of wall and independent columns constitutes the vertical structure, providing flexibility for spatial and landscape organization. Hence, the space inside the gallery conveys a feeling of flowing movement and diversity. The subtle relationship between beams and slabs, where beams only appear on the periphery without crossing into the gallery, makes the roof seem weightless. The piece is positioned along the lakeshore with its platform and staircase cantilevering into the lake to further emphasize the floating image. As for the decorative design, Xiang used a slender concrete grille, simplified hanging and other techniques. For example, the planting bed is designed with an extremely interesting structural intention to make it appear as if it is also floating.

This piece expresses many characteristics of early modernist architecture, but also reflects the spatial interest of Chinese garden making. If the Sun Yat-sen Mausoleum Music stage in the 1930s is an exemplary landscape design that combines Chinese and Western classical architecture, then the Yushanhu Park piece is a model of regional exploration with modernist architectural concepts in the 1970s. Despite the limitation of construction technique, Prof. Xiang was still able to focus his practice more on the exploration of improved architectural design, giving his works a long-lasting resonance.

Zheng Yong
Chief Architect, Specific Architects
Doctor of Architecture

南侧山墙 South view

# Fuyuan Market, Ma'anshan



1984 / 1985

# 马鞍山富园贸易市场

1980年前后，项秉仁开始关注建筑语言、表征及符号的研究，同时也慢慢建构起建筑同环境、行为和心理模式之间相互关系的概念，希望能在创作实践中进行大胆探索和尝试。正值1984年至1985年间，作为博士学位研究课题的一个组成部分，项秉仁参加了安徽省马鞍山市富园贸易市场的规划设计工作。

富园贸易市场坐落于马鞍山市中心区，距离景色秀丽的雨山湖不远，是用作当地农副产品销售的建筑群。基地上现有的建筑给规划带来了一定的限制，但也成为这个建筑群的固有特色。考虑到原有的粮食加工厂对外交通的需求，项秉仁将基地分为两个部分，依据形状分别安排了小商品市场步行街和农副产品大棚。

在当时这个项目设计中，项秉仁实践了一些创新的理念：其一是关于城市设计片段的探索。以中国江南传统城镇空间特征的街道、广场等外部空间为设计主题，建筑物则作为空间的围合界面而存在。商业步行街和广场空间当时在国内还很少见，不仅化解了地形狭窄的劣势，同时也将城市人流引入建筑群内部，巧妙地化被动为主动。其二是关于现代和传统的对立统一。这是对文丘里《建筑的复杂性与矛盾性》中提出的"兼容的困难的统一，而非排斥的简单的一致"的实践。现代建筑的设计手法插入到传统空间格局之中，"打破方盒子，把透视的三度分解为二度的面，再以非透视方盒子的形态重新构成起来"，运用建筑符号学的观念在建筑立面造型、细部以及构件上对传统元素进行变形和裂变以体现时代精神。

这个设计在1980年代出现时，具有一种以现代建筑特征体现中国传统文化内涵的新建筑形象。它不仅应和了当时国内外规划和建筑设计界的后现代主义思潮，也因其注重空间特征和现代风格的努力而在当时赢得了王明贤等国内著名建筑评论家的广泛关注和正面评价。"中国未来的建筑不可能都是高标准、新材料，但只要我们注意到建筑的布局，新与老的联系、城市空间的设计、基地表面，水和绿化的处理、小品建筑的设计，注意到人际交往和领域性的需要，注意到文化内涵和意义传播，我们仍有可能创造出一流的建成环境。"

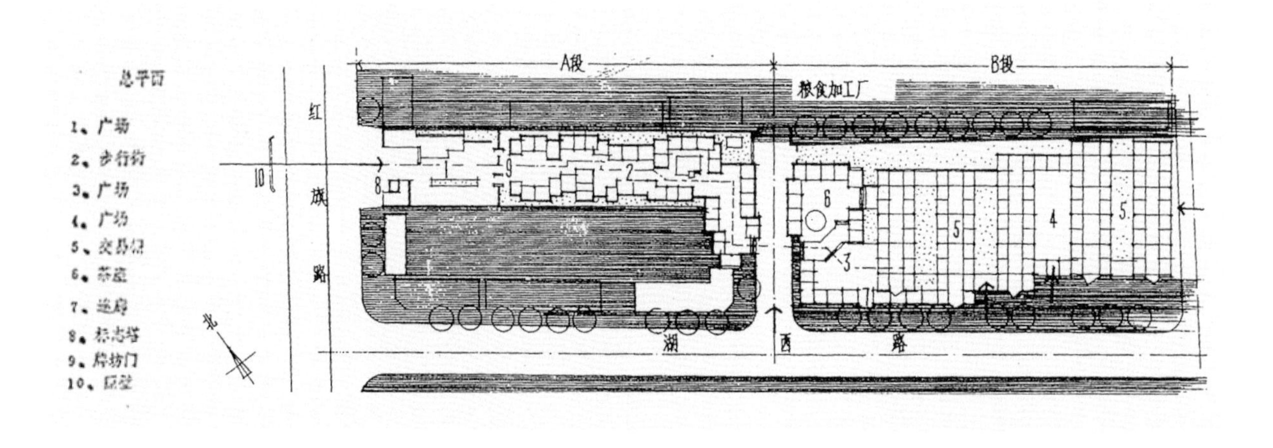


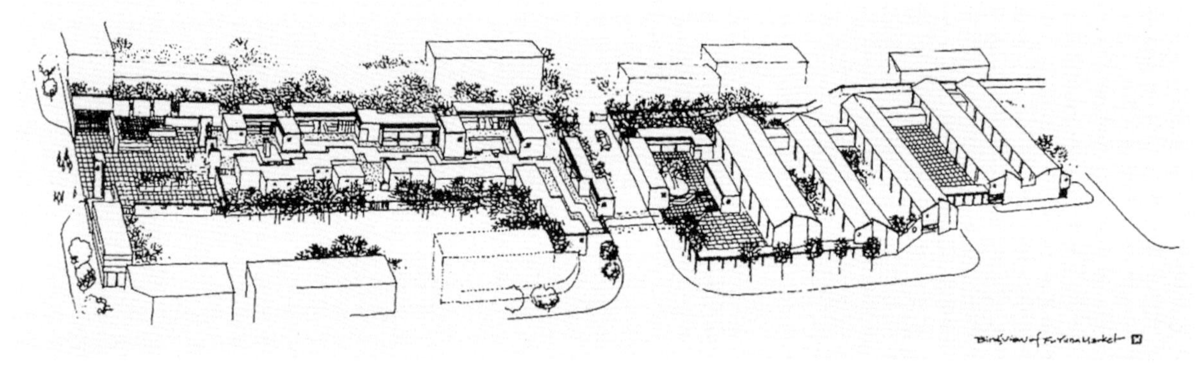

︿ 总平面图 Site plan
﹀ 市场全景鸟瞰 Bird's-eyes view

Around the 1980s, Xiang Bingren began to find a way to relate architecture, built environment with behavioral and psychological patterns. After extensive reading and learning of architectural theories, both domestic and foreign, he was desiderating an opportunity to explore these theories in practice. At the time from 1984 to 1985, his moment came. Xiang Bingren participated in the planning and designing of the new Fuyuan Market at Ma'anshan City as part of his doctoral research project.

The site of Fuyuan Market, located in central Ma'anshan City, is not far from the picturesque Yushanhu Lake. The project, an architectural complex, was designed and built to accommodate the selling of local agricultural by-products. Existing construction on the site imposed limitations while providing characteristics for the planning and design of the market complex. Xiang Bingren divided the site into two parts, due to the grain processing plant's need for exterior accessibility. He arranged sheds for the warehousing and sale of farm products in addition to a pedestrian street which would serve as a small, site-specific commodity market.

Many of Xiang Bingren's courageous innovations and conceptual explorations are highly visible in this project: 1) The exploration of how to design a piece of urban fabric. The project redeployed the signature street space of traditional towns in southern Yangtze River delta and focused on the design of outdoor spaces such as plazas while using

# Fuyuan Market, Ma'anshan

市场标志 Market landmark

buildings as instruments for enclosing spaces. At that time in China, the notion of designing commercial pedestrian streets and plazas was a relatively new one, but, in this project, the architect quite effectively solved the problem caused by the rather linear and narrowly shaped site and successfully guided the stream of people into the complex. 2) The exploration of the conflict and unity between modernity and tradition. The project practiced Robert Venturi's ideal that "(architecture) must embody the difficult unity of inclusion rather than the easy unity of exclusion" in *Complexity and Contradiction in Architecture* by integrating modern design methods with traditional spatial structure. The architect also borrowed from the emerging inclination to "deconstruct the box, break down three-dimensional perspective into two-dimensional plates, then reconstruct the box in a non-perspective form". The notion of architectural semiology was applied to the design of facades, details and building components, where the transformation and fission of traditional elements reflect the spirit of a new era.

As a preliminary attempt of design exploration, when this project debuted in the 1980s it presented a "new look", which was mostly modern yet bore the unmistakable character of traditional Chinese culture. This project not only echoed the postmodernist trend that was prevailing in the architectural academia but also contained the modernist's space-oriented efforts. Hence, the project was highly regarded and

入口牌楼 Main entrance

gained extensive, nationwide attention from architectural critics such as Wang Mingxian. "For future constructions in China, we cannot treat each and every one of them with the highest standard while applying the most advanced material. However, as long as we carefully deal with the layout of buildings, the relationship between old and new, the treatment of hardscapes/softscapes and architectural pieces, the design of urban space while paying attention to the needs of interpersonal communication and territoriality as well as cultural connotation and conveying of meanings, we can still create world-class built environment."

## 评论
## Review

他对环境行为的心理问题的关注，对建筑符号、象征和意义的研究，对人的历史文化传统的考虑，使这一建筑群成为有意义的环境。他在设计里成功地运用了符号学“语义裂变”的手法，如大门设计考虑到人们的风俗爱好，采用了传统的四柱三间三楼的原始构图，但完全没有受到传统做法的束缚，大门成为一个二度板块的构成。[1]

王明贤

著名建筑评论家，中国“实验建筑”倡导者

与当下宏大的商业体相比，规模小巧的富园贸易市场不足挂齿，却是改革开放初期最早综合运用城市设计方法、环境行为学、后现代主义等理论进行创作的重要实践之一，也集合了项先生当时在博士阶段的多方位思考和研究成果。在略带传统韵味的质朴外表下，我们也许更应该关注的是建筑师的理性探索，这些方面包括适宜的设计策略、城市空间关系的巧妙应对、对江南城镇街巷空间特征的挖掘以及具有人文关怀的城市公共空间营造，今天看来仍然具有其积极的意义。

陈强

同济大学建筑城规学院硕士生导师
上海道辰建筑师事务所（DCA）主持建筑师

His keen interest in environmental and behavioral psychology, research on architectural symbolism, tokens and meaning as well as his consideration of human's historical and cultural traditions have made this building complex a physical environment with strong cultural implications. He successfully implemented the methods of "semantic fission", originating from semiology, in the designing process of this project, such as the design of the archway where he assimilated local customs and preferences into consideration by adopting the traditional "four columns, three gates, three brims" composition but he also was not constrained by the tradition, making the archway part of the "two-dimensional plates".[1]

Wang Mingxian

Renowned architectural critic, Campaigner of "experimental architecture" in contemporary China

While it might seem insignificant compared with all the enormous commercial buildings which receive our attention, it is these small-scale markets, such as that at Fuyuan, that held extraordinary importance during the early stages of China's reform and opening up. Such community spaces integrated urban design methods, environmental behavior study, postmodernism and other theories. This piece also epitomizes Mr. Xiang's multi-faceted thinking and research for his doctoral study. Beneath the simple yet traditional appearance of this piece lies the rational explorations of the architect: finding appropriate design strategies, responding to urban spatial relationships, exploring the features of many towns and streets surrounding the Jiangnan area, and creating urban public spaces that on a human scale. All of these efforts still hold positive significance today.

Chen Qiang

Master Supervisor, CAUP, Tongji University
Chief Architect, DCA Studio

1 王明贤：《新建筑的理想与宇宙图案》，《南方建筑》1993 年第 1 期。
Wang Mingxian,"Ideal of New Building and Cosmic Pattern"(in Chinese), *South Architecture*, no.1(1993).

# Luyuan Market, Kunshan

1986 / 1987

# 昆山鹿苑市场

在完成富园贸易市场一年之后，项秉仁又获得了一个可以实践城镇形态环境符号学的项目——江苏昆山市鹿苑市场。项目坐落于昆山市中心的娄江河岸，基地没有直接临街，而是处于现存的两幢形式迥异的多层楼房背后。在20世纪80年代，那里是当地的小商品和农副产品的集散地，街道景观缺乏统一规划，致使沿街建筑的体量、式样和色彩呈现出较为混乱的城市形态。

项秉仁认识到这个项目的设计不只是创建一个普通的专业市场，而且是创造一处新的城镇形态环境，同时由于基地临江的特点，需要一并处理好沿江景观与现存建筑的关系，使其能巧妙地组织进新的市场建筑环境之中，成为一个有机的整体。整个建筑群的布局仍然以传统城镇的城市街道和广场空间形态为空间原型，并通过建筑形体、细部处理、街市小品等来传承传统地方文脉。在赋予现代城市空间几何性的同时，建筑物继续作为界定城市空间形态的元素而存在。外观和细部保持了中国城市民居以灰白色调为主的朴素、统一和整体的效果，间或点缀的鲜艳色彩赋予建筑时代感。与此同时，建筑物运用适当的材料技术，以低廉成本在短期内建成并投入使用。

为进一步表达设计理念和创新想法，项秉仁用水粉画的方式绘制了它的效果图，这张建筑轴测图清晰地表达了广场和街道城市空间的组成，画面光影效果丰富、生动且富有生活气息，典型的印象派风格增强了画面的韵味。

城镇形体环境蕴含着复杂多样的含义，是一种特殊的符号系统。在城市形态的演进中，一部分旧的形体被赋予新的意义，而新的形体也不同程度地沉淀旧的脉络，城市形态和建筑在这种多样化的历史进程和环境更替中向人们传播着丰富的历史文化内涵。鹿苑市场设计清晰地表达出：城镇建设对传统文化的继承应该更多反映在精神内涵而非形式化的模仿。从富园贸易市场到鹿苑市场的设计过程反映了项秉仁在不同地点、不同的对象条件下，对于建筑环境及其文化内涵的思考。

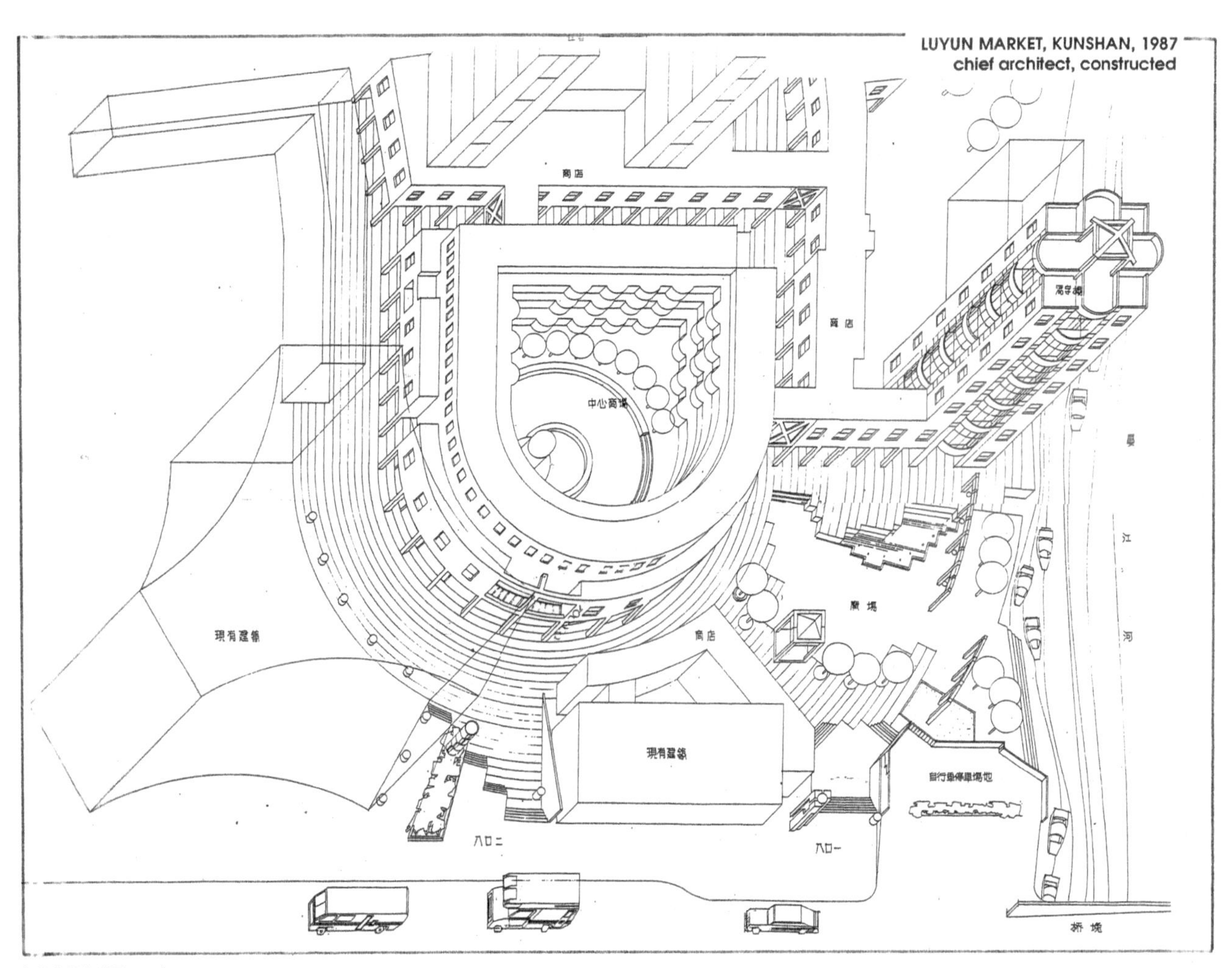


市场全景鸟瞰图 Bird's-eyes view

One year after the completion of Fuyuan Market, in 1986, Xiang Bingren was commissioned with the design of Luyuan Market in Kunshan City, Jiangsu Province, which granted him another opportunity to apply his study of urban physical environmental semiology. The site of this project was on the bank of the Loujiang River that runs through central Kunshan. The site was not actually along the street but at the rear of two existing multi-story buildings with different architectural styles. During the 1980s, that neighborhood was crowded with peddlers selling small goods and farm products. There was no overall planning for the streetscape and street frontages were defined by buildings of various styles, volumes and colors presenting a chaotic scene as a backdrop for bustling crowds.

Xiang Bingren realized that this project was less about building a conventional market and more about creating a new type of urban physical environment. Because it was adjacent to the river, he understood that he must deal with the riverside scene and existing building volumes carefully while subtly combining these elements with the new market in pursuit of a single, unified aesthetic. The layout of this building group was decided based on the traditional form of blocks and streets, reflecting local context through carefully designed building volume, details and street furniture. For this design, buildings are not only about geometry, but also contribute to defining urban spatial form. The dominant tone of grey and white on facades and details of the building maintain the simplicity, unity and whole-

室内空间 Market interior

ness of traditional Chinese dwellings while his inclusion of bright colors in accented continuity conveys modernity. At the same time, appropriate techniques and materials were applied, facilitating the low-cost construction within limited time which reflected his prevailing postmodernist values.

To fully present his design ideal and innovative thinking, Xiang Bingren did the rendering of the project personally in the format of a gouache painting. That axonometric drawing was very vivid and full of life. It effectively showed the composition of plazas and streets as well as their dynamic shadow and light. The classic impressionist style of the drawing enhanced the postmodernist charm of the design.

Urban environments are composed of layers of meaning in a continuous, multi-valiant semiotic system. Throughout the evolution of urban form, many old physical entities are gradually endowed with new significance while some new physical entities partially inherit meanings from the past. In the course of history, through constant changes, the city's physical environment and buildings narrate the rich historical and cultural connotations. While conscious of the traditional culture it has inherited, Luyuan Market is more committed to being alike in spirit rather than pursuing similarity of form. Xiang Bingren enhanced his understanding of built environments and their cultural connotation through the design of both the Fuyuan and Luyuan Markets, despite different locations and contexts.

^ 局部细节 Details
> 沿街立面 Street view

# 评论
Review

1986年，正在翻译《城市的印象》的项秉仁先生，开始着手设计鹿苑市场。从那张漂亮的轴测效果图，可以看到日后成熟期的项秉仁，以及他所完成的宁波文化广场和招商局上海中心的影子。城市成为建筑设计的关键要素，建筑语言追求的是典雅和高技的融合。

鹿苑市场的空间主题从城市出发，有圆形的古希腊剧场、长方形的罗马广场、引导性的街道空间、控制边界的桥。城市在建筑里投射出丰富性与复杂性。建筑的语言是典雅和含蓄的，大厅之上的螺旋楼梯和市场两端的桥梁，是两种不同节奏的联系；蓝红两色方向不同的桁架，述说着空间的引入和导出，手法简朴却保持统一。

作为项秉仁先生的早期作品，鹿苑市场已经为未来的发展种下了一粒“种子”，从城市角度来思考建筑空间，用历史和未来的双视角关注建构手法，这是一段美好历程的开始，精彩至此展开。

刘江

上海经纬建筑规划设计研究院股份有限公司副总建筑师
建筑学博士

In 1986, Mr. Xiang Bingren articulated the design process of the Luyuan Market while he was translating *The Image of the City*. A beautiful axonometric rendering of the market revealed similarities shared between some of his works from more mature stage of his practice, such as the Ningbo Cultural Plaza and the China Merchants Shanghai Center. All these works uphold the city as a key element for architectural design while showing how architectural language might pursue a fusion of elegance and integrated technology. The design of Luyuan Market took off from a spatial theme derived from the city. In the design are included strong suggestions of a semi-circular ancient Greek theater, a rectangular Roman square, manipulative street spaces, and a bridge that sets the boundary, where the city projects richness and complexity in architecture. The methods adopted for the design are simple yet unified with an architectural language that is subtle yet elegant: a staircase spirals above the hall while bridges at both ends of the market create a connection between two different rhythms. Trusses stretch in competing directions and are painted in blue and red depicting the experiences of entering and exiting space itself.

As one of his early projects, Luyuan Market planted a “seed” for Mr. Xiang’s future career path. It was here that he began considering architectural spatial design from the perspective of the city context while also contemplating architectonic possibilities with both the past and future in mind. This was the beginning of a beautiful journey, from which wonder unfolded.

Liu Jiang

Deputy Chief Architect, Shanghai Longilat Architectural Design & Research Institute Co., Ltd.
Doctor of Architecture

模型鸟瞰 Bird's-eyes view

# Hu Qing Yu Tang Pharmaceutical Tourist Area



1987 /

# 胡庆余堂药业旅游区规划

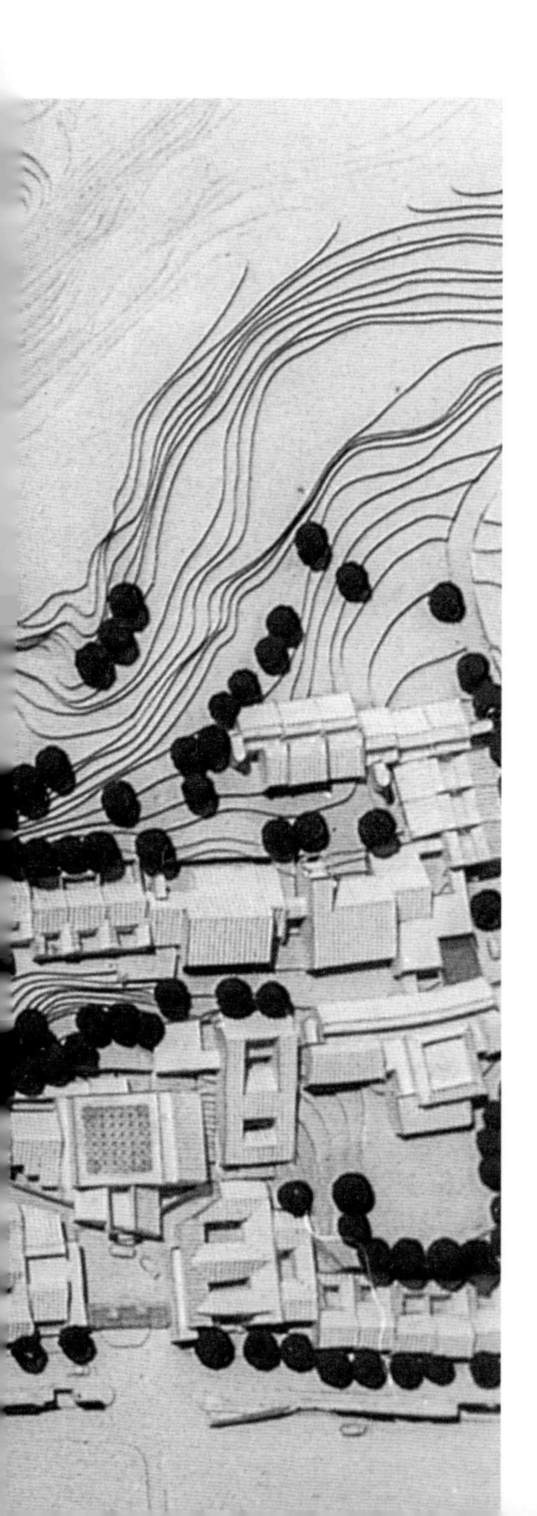

坐落于杭州吴山山麓的传统江南民居建筑群中，胡庆余堂作为江南首屈一指的中药堂享誉中外，拥有将近150年的历史，其建筑也因其典型优美的古徽州风格被列为受保护的历史文物。项秉仁在胡庆余堂药业旅游区规划中尝试以胡庆余堂为主要特色建筑，延展其功能和文化含义的同时，开辟将中药加工、销售、陈列、医疗、培训和旅游融为一体的综合旅游区。

旅游区的空间叙事围绕着“传统形态”与“现代功能”展开。以不破坏建筑群的原有文化内核为目的，尽量保留历史遗留街道的空间尺度和广场形态；适应现代生活的小商品业态注入到被保留的原有旧城纹理建筑之中。传统建筑的屋顶特色、建筑高度、体量、结构、色彩等都得到了充分的展现，即使是新建的较大规模设施，也尽量化整为零，与现有城市文脉相协调。

胡庆余堂药业旅游区规划与富园贸易市场、昆山鹿苑市场同属于项秉仁早期的三个城市建筑设计，也都反映了那段时期他在后现代主义思想影响下所采取的设计态度。该规划是项秉仁对于中国国内城市规划现状的首次接触与探索，同时也因有幸与同济建筑学院的同事们合作而学习到冯纪忠教授的专业见解：在努力保护历史城市街区的街巷路网空间形态以及传统建筑群体的屋顶肌理的前提下，合理地引入现代生活所需的功能。这是项秉仁对旧城历史保护与更新的初步尝试，体现了他注重街巷、广场与建筑原始形态的理念，为之后的杭州中山中路和元福巷等保护与更新奠定了基础。

---

Located in the piedmont of Hangzhou's Wu Hill in a traditional southern Yangtze River settlement, Hu Qing Yu Pharmacy is the most prestigious Chinese medicine shop in the Yangtze River Delta. The pharmacy itself is also listed as an official cultural heritage site because of its typical and beautiful ancient Huizhou architectural style. Hence, in the planning of the tourist area, Xiang Bingren attempted to expand the cultural and functional connotation of Hu Qing Yu Pharmacy. He made it the main

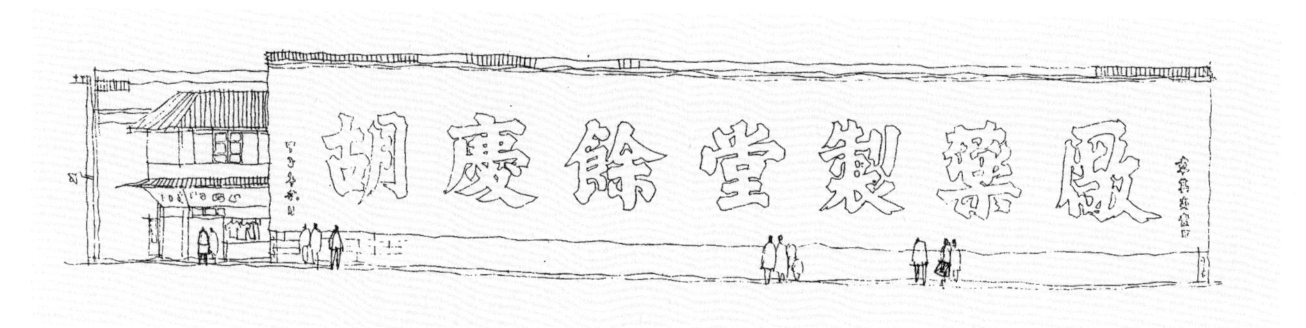


手绘沿街立面 Street view sketches

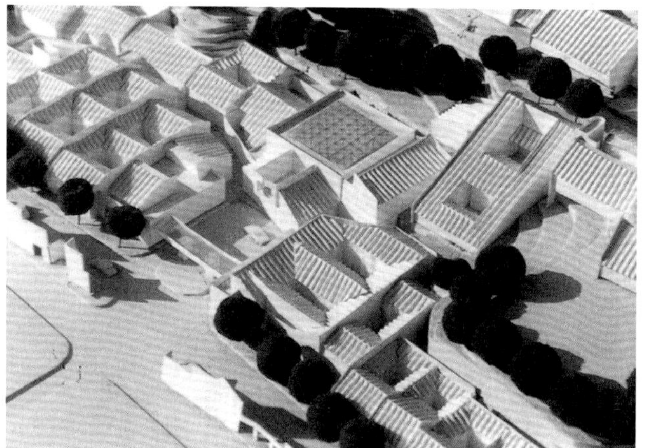

模型局部 Model

feature of this tourist area while developing a comprehensive identity that integrates tourism with the processing, displaying, and trading of Chinese herbs and medical treatment.

The spatial narrative of the tourist area is focused primarily on "traditional forms" and "modern function". The cultural core of the settlement was preserved while the scale and form of the historic streets and plazas were kept intact. To meet the needs of modern tourists, small businesses were set up with an arrangement and aesthetic meant to preserve the historic town fabric. The design paid due attention to the traditional roof characters, building height, volume, structure and colors of the building. During the construction phase of the large scale facilities, the volumes were strategically broken down to accommodate and blend into the existing context.

Xiang Bingren's early career included the planning of the Hu Qing Yu Pharmaceutical Tourist Area, the design of Fuyuan Market and the Luyuan Market. In their own ways, they represent his eclectic design

# Hu Qing Yu Tang Pharmaceutical Tourist Area

模型局部 Model

philosophy which was heavily influenced by the prevailing postmodernism of that time. This project was the first time that Xiang Bingren got involved in urban planning practice in China. His career and practice were furthered by his collaboration with a design team from Tongji University where he had the honor to learn from Prof. Feng Jizhong. It was under Feng's guidance that Xiang Bingren gained the professional insight that urban planning should spare no effort to protect the figure-ground layout of the historic areas while adapting them to modern life. This project is important for a variety of reasons. Firstly, it was the beginning of Xiang Bingren's persistent exploration on the preservation and regeneration of old cities; secondly it demonstrated his budding concerns for streets and alleys, of plazas and primitive form of architecture; thirdly, it equipped him with a better understanding of preservation and renewal for a similar project he would encounter later—Mid. Zhongshan Road and Yuanfuxiang.

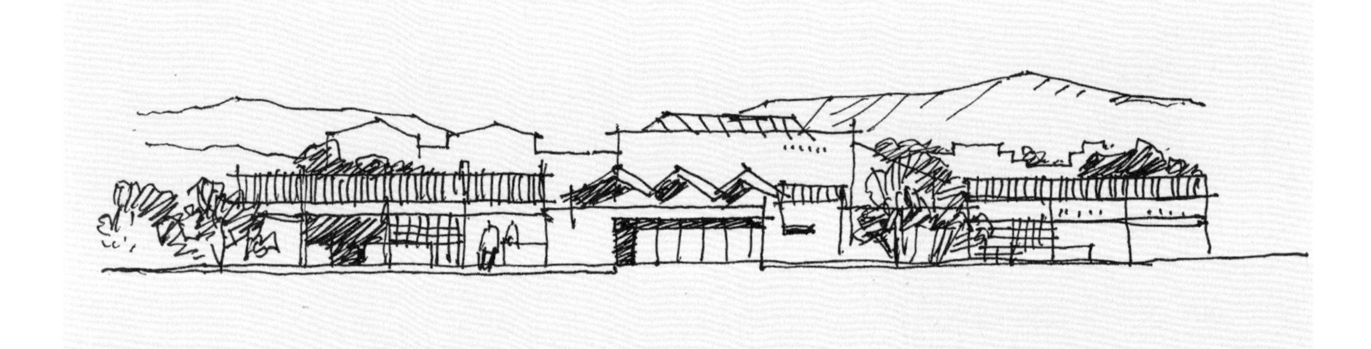

手绘草图 Freehand sketches

## 评论
## Review

本项目展现出项秉仁先生一贯的设计态度，即在诸多矛盾的权衡中探求具有说服力的解决答案：尊重传统文化的同时，响应当下社会的发展需求；延续城市风貌的同时，以一种鲜明的时代姿态与环境对话；驾驭新技术新材料的过程中，细部设计融入文化艺术的痕迹。先生面对"传承与发展""地域性与世界性"等宏大社会命题时，试图通过建筑设计的理论、策略和技巧，构架一座桥梁，建立一种平衡。

祁涛
深圳华森设计执行总建筑师
建筑学博士

This project also reveals many of the tensions inherent in Mr. Xiang's design attitude. In all his projects he seeks to find a persuasive solution to a variety of challenges while balancing many contradictory aspects: traditional culture versus development needs of the current society; the overall urban image verses the buildings' unique qualities; new technologies and new materials versus cultural and artistic considerations. In the face of grand social propositions such as "inheritance and development" and "regionality and cosmopolitanism", Mr. Xiang has always attempted to build a bridge and establish a balance utilizing theories, strategies and skills of architectural design.

Qi Tao
Executive Chief Architect, HS Architects Shenzhen
Doctor of Architecture

# 2

## 上海复兴公园园门重建
Fuxing Park Gate Renovation

## 雨花台南大门建筑
South Entrance of Yuhuatai Memorial Park

## 杭州中山中路历史街区保护与更新
Protection and Renovation for the Historic Neighborhood of Mid. Zhongshan Rd., Hangzhou

## 杭州元福巷历史街区保护更新
Conservation and Renovation for the Yuanfuxiang Historic Block, Hangzhou

## 城市历史保护与更新：尊重与创新

对于城市形态和城市设计的重视，对于城市建筑文化遗产的尊重和延续，对于建筑的当代性和创新性的追求一直贯穿于项秉仁所接受的大小设计项目委托之中，上海复兴公园大门建筑设计（2000）和南京雨花台南大门建筑设计（2001）就是其中的两例。鉴于当时国内对于城市设计认识的普遍偏颇，项秉仁试图以城市的视角去构思建筑设计并通过把握形式风格、空间尺度、材料质感、细部处理等方面去处理好每一件建筑物与建筑小品来最终实现城市设计的目标。

随着城市化进程的加速，在城市建设的发展过程中，如何处理好城市历史街区和历史建筑的保护和更新问题是项秉仁在建筑实践中需要直面的。城市设计策略的制定不仅是为后续的实际保护和更新建设提供守护边界，更为激发街区的历史人文价值，可持续地为街区空间注入现代产业活力提供指引。

## Historic Preservation and Urban Renewal: Respect and Innovation

In all of his design projects, large or small, Xiang Bingren always values the existing urban form and its renovation. He respects the architectural heritage of the city and tries to keep it in an innovative way; he seeks for both contemporaneity and originality in his design. Two exemplary cases are Fuxing Park Gate Renovation (2000) and South Entrance of Yuhuatai Memorial Park (2001). In view of the general biased understanding of urban design in China at that time, Xiang tried to conceive architectural design from the viewpoint of the city, and managed to fulfill his ambition in urban design by treating each building and architectural feature in terms of formal style, spatial scale, material texture, and details.

The progress of urbanization raises a question for Xiang and indeed all the architects and city planners: how to deal with the protection and renewal of historical blocks and buildings in the city? Urban design strategies should not only be the guidelines for subsequent conservation and renewal actions, but also operate like the stimulus for historical and cultural values, and sustainably inject modern vitality into the place.

入口立面效果图 Rendering of the entrance

# Fuxing Park Gate Renovation



2000 / 2000

# 上海复兴公园园门重建

复兴公园历史悠久，一个多世纪以前，这里曾是一片农田、村舍，称顾家宅。1900年八国联军攻陷北京，法军侵入上海，在顾家宅建立起养马棚和屯兵地。1901年法公董局将此地开辟为法国公园，于1909年6月建成开放，专供法国侨民游乐休闲。第二次世界大战期间日军曾一度占领此地，在公园内设立兵营。1943年汪伪政府接管后称其为大兴公园。直至1945年抗日战争胜利始更名为复兴公园。

2000年复兴公园园门重建计划是在当时上海正处于努力改善城市绿化环境的形势下进行的。此前，复兴公园重庆南路的园门也曾经历过多次重建。项秉仁接受这个项目计划的设计委托也有一些渊源：一方面，他青少年时期的居住环境和背景使他对复兴公园有种特别的情感；另一方面，当时城市设计的概念正慢慢兴起，学术界涌现了许多大型城市空间的漂亮图纸和模型，却鲜有真正品质良好的城市小环境。他希望能通过这样一个城市小品建筑，运用城市的观念去构思建筑设计和环境营造，用专业的素养把握建筑的形式风格、空间尺度、感觉、材料质感和细部等，实现城市设计真正的目标。

"数十年前的复兴公园园门很简单，是一榀木制的花式大门，具有法国乡村风情。这是百年历史的见证，也是区别于其他城市公园的最大特色。如果新的园门能沿用这样一款原真性的大门设计，定可以保持和延续这种历史感并唤起人们对当时历史记忆的联想。同时，这一'历史残片'还应镶嵌在一个现代的形式之中，符合当代建筑美学精神。两者之间能形成某种强烈对比又蕴含深意的新建筑造型。"所幸的是，团队在成堆的历史档案中翻查到了1932年顾家宅公园的木制园门的原始设计图纸并予以复原，使新设计得以从一个真实的历史依据展开。

木质花式大门宽9米，分成4扇，其两侧为干挂白色水晶花岗石板的售票亭建筑，形式与大门的设计风格相似，均带有装饰派的韵味，以摹拟"历史残片"。两片体现现代建筑特色的清水混凝土墙从两个售票亭的外侧下部"贯穿"至内侧，再与木质大门的立柱连接，造成"历史残片"的镶嵌感。为了同时形成椭圆形广场的空间围合和公园内外视觉的通透，清水混凝土墙被分成三段，并用水平向的扁钢围栏相连接。两片混凝土墙，分别立于椭圆型广场的两端，使广场与人行道的转角处有一个明确的建筑处理，并向外延伸，成为新建的公园围栏。

法式乡村风情的木制花式大门 Wooden door with French rural style

为了呈现更好的城市小空间环境，建筑细部被反复推敲：大门前广场的大小尺寸、园门的宽度高度以及售票亭的体量和高度与人体的尺度关系；围栏的尺度和通透度；清水混凝土墙体的分块尺度，分缝的处理；扁钢栏杆的选料、排列间距和油漆色彩；售票亭外墙干挂的水晶花岗石板的分块和分缝；门窗选料和色彩的决定乃至广场铺地石料的设计、选材、分缝等。

今天看来，当时的中国城市设计正逐步从纸本绘制进入更重视城市环境的新阶段。每一项工程，无论规模大小、造价高低，都是属于城市环境的细胞个体，其品质的优劣直接影响着城市的整体形象。

Fuxing Park, an expansive and complex space with a rather long history in central Shanghai, was previously farmland with a few village houses called Koukaza. On August 14, 1900, when the Eight-Power Allied Forces occupied Beijing, the French army invaded Shanghai to establish a horse range and squadron in Koukaza. In 1901, the French Public Affairs Bureau converted this area into the French park, which opened in June 1909 for accommodating the recreational impulses of French nationals. During the Second World War, it served the Japanese army as barracks. After the takeover of the Wang Puppet Government in 1943, it was renamed Daxing Park and then finally as Fuxing Park when the War of Resistance Against Japanese Aggression ended in 1945.

In 2000, the renovation plan of Fuxing Park Gate was carried out as part of the Shanghai municipal government's efforts to improve city

Project

# Fuxing Park Gate Renovation

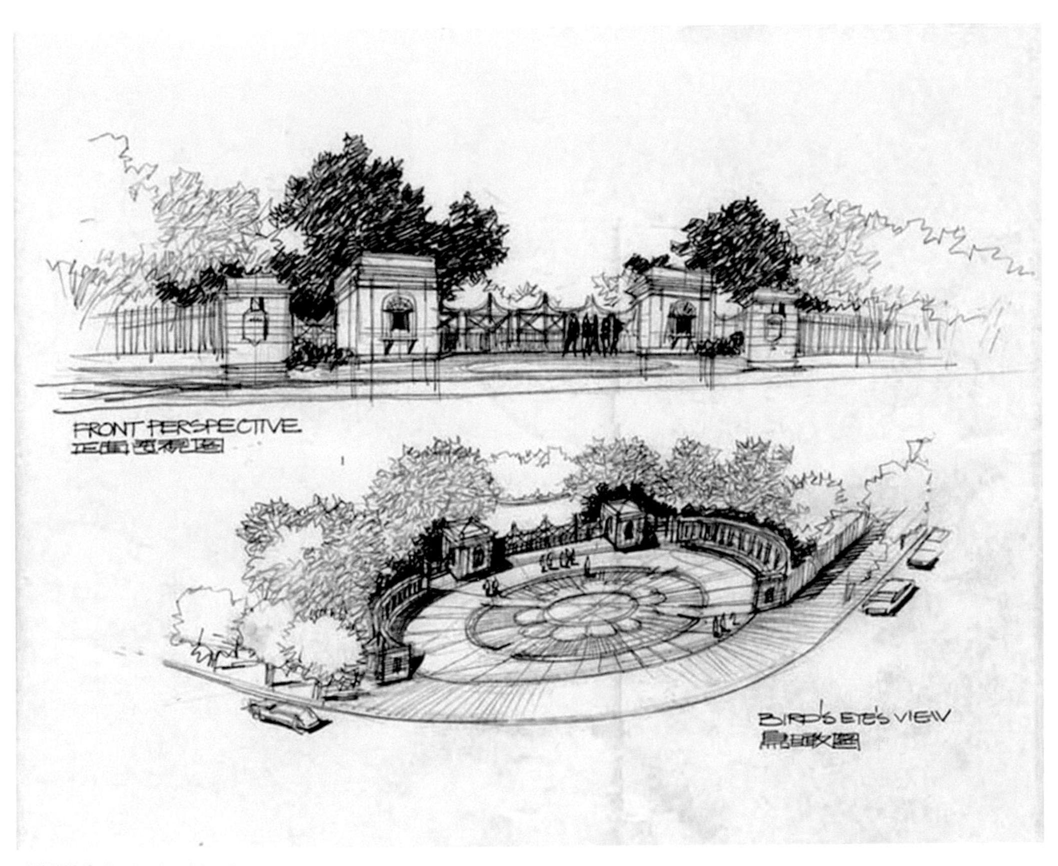


手绘设计稿 Hand-painted sketch

afforestation. Previously, the gate on South Chongqing Road had been through several rounds of renovations. Xiang Bingren's involvement in this project seems, in retrospect, somehow destined: first of all, his childhood experience in the neighborhood made him emotionally connected to Fuxing Park; secondly, although urban design as a concept was emerging and many intricate drawings and models were made in the academia of that time, few small-scale urban places had been actually designed. Therefore, Mr. Xiang hoped to bring out the true value of urban design through such small-scale urban pieces, where he could explore architectural design and place in the view of urban environment, dealing with form, style, scale, experience, material and details professionally.

Decades ago, the design of Fuxing Park Gate was quite simple: a wooden French-style single-leaf gate, representing its history over a century and which distinguished Fuxing Park from other parks in the city. If the new gate could bring back the essence of the original one, it could certainly reinforce the historical continuity and evoke urban memories. However, this "piece of the history" should be integrated with a modern form, in line with the contemporary aesthetic spirit of architectural design. The drastic contrast between old and new could create a form of profound meaning. Fortunately, they found original design documents of the gate of Koukaza Park in 1932 from piles of historical archives, providing the design of the new gate with some reliable historical basis.

The wooden gate is 9m wide, divided into 4 door leaves, with two ticket booths flanking either side. These booths are covered with white crystal granite slabs. Similar to the design of the gates, they also bear decorative charm to simulate the design concept of "historical frag-

^ 钢百叶围栏 Metal fence
v 门墩 Gate pier

^ 广场雕塑 Sculpture in square
v 墙面细部 Details of wall

ments". Two pieces of modernist clear-water concrete wall "penetrate" from the outside of the two ticket booths to the inside, connecting the pillars of the wooden gates, creating an image of "inlaid historical fragments". In order to emphasize the spatial enclosure of the elliptical entrance plaza while promoting the visual connectivity of the inside and outside of the park, the concrete walls are divided into three sections and connected by a horizontal modern flat-steel fence. The two concrete walls are placed on either end of the elliptical plaza to reinforce the street corner while their outward extensions serve as fences of the park.

In order to create a better micro-urban space, Mr. Xiang repeatedly and carefully scrutinized some design details. In terms of scale, he carefully worked out the size of the square in front of the gate, the width of the gate itself, the proportion and transparency of the fence, the correlation between the dimensions of the ticket booth and the scale of human body. Materially speaking, he took special care of the subdivision and cleavages of the clear-water concrete wall, the material for the flat steel railing, and its spacing and paint color, the subdivision of the hanging crystal granite slab on the outer wall of ticket booth, and the stone paving. He even handpicked the color and material of the door frames and window frames.

Nowadays, urban design in China is no longer simply about drawings on paper but gradually moving toward a stage that places more emphasis on the actual urban environment. Every project, regardless of its size and cost, is considered as a living cell of the built environment, whose quality directly affects the overall image of the city.

## 评论
## Review

当我们回顾项先生在本世纪初的几个城市"小"设计的时候，都会惊讶于他对城市微更新的积极践行和精益求精的态度。彼时的学界，城市设计的概念正渐渐兴起并备受关注，然而关于城市设计的研究和实践却并不广泛。城市设计的理想最终实现，不仅需要城市建设领导者的高瞻远瞩，施工人员的匠心实施，更需要有大批有素养、具备专业水准的规划师和建筑师去实践那些被深思熟虑、细心推敲的设计方案。

在项先生的这个作品里，我们看到了他自觉的城市意识。用城市的观念去构思建筑设计，用历史的要素去回溯城市文脉，用当下的建筑语言去体现城市时代精神，并细致关注建筑尺度、材料质感、细部处理等建构问题。看似只是城市中的点点滴滴，却体现了他对实现美好城市的憧憬。

滕露莹

上海秉仁建筑师事务所合伙人、副总建筑师

Looking back on Mr. Xiang's several "small" urban interventions at the beginning of this century, it is easy to be amazed at his active engagement and pursuit of excellence in micro-urban-renewal. At that time, urban design received more and more attention as an emerging concept, without extensive research or practice. The implementation of urban design requires not only the forward-looking vision of city leaders and the ingenuity of engineers. It also requires a large number of well-educated professional planners and architects to execute those thoughtful and carefully designed schemes.

In this work, we see Mr. Xiang's urban consciousness: conceiving architectural design with an urban context in mind, tracing urban history with key elements, emphasizing the contemporary spirit of our cities with a modern architectural language. He also pays close attention to architectonic problems such as scale, material texture, and detail design. These small interventions seem trivial to a city, but they authentically and powerfully reflect Mr. Xiang's longing for a beautiful city.

Teng Luying

Partner/ Deputy Chief Archicect, DDB Architects Shanghai

大门入口 Entrance

# South Entrance of Yuhuatai Memorial Park



2001 / 2001

# 雨花台南大门建筑

南京雨花台南大门设计是继上海复兴公园园门之后又一个通过建筑呼应环境、环境烘托建筑去塑造城市小环境，从而最终实现城市设计目标的尝试。基地位于南京市雨花台区行政中心的对面，正对十字交叉的环岛，在一片郁郁葱葱的密林之间。南大门作为南向景区轴线的起始点，场地空旷舒展，尺度很大。众所周知，南京雨花台是新民主主义革命烈士殉难处，有着浓烈的红色纪念氛围，其北大门的设计也因此格外庄重肃穆；而事实上它也同样是南京著名的风景旅游区，需要体现其公共性和开放性。

建筑形体采用了极其现代简约的语言，在150米宽的城市界面上设置了四个体块元素——两个售票亭、一个值班室和一个小卖部。这四个功能不同的建筑小体块将南大门建筑在水平向划分为3段，建筑与围栏交替出现，着意表达领域的划分而弱化其封闭性及礼仪性，并与周边环境融为一体。在简洁的形体基础上，项秉仁对材料和细部有特别的刻画。建筑材料选用石材与玻璃来加强虚实对比并烘托建筑的历史感与现代感。立面的花岗石划分采用宽窄不同的水平线条镂空处理，在厚重中突显出细腻与舒展。大门两端自由镶嵌着许多圆形的石雕光纤灯，象征"落花如雨，雨落石花"的寓意，充分体现了南京城市的创新性和当代性。简洁有力的形体组合，开放灵活的空间界面，别具匠心的立面细部，令建筑本身仿若生长在雨花台之上，与城市相契合。

The design of the south entrance of Nanjing Yuhuatai Memorial Park is another attempt to achieve urban design goals through the construction of context-responsive buildings after the renovation of Fuxing Park Gate. The project is located opposite of the political center of Yuhuatai District in Nanjing, facing the roundabout at the crossroads among a lush jungle. The South Gate, as the starting point of the southern axis of the park, opens into stretches of space with a large scale. The Nanjing Yuhuatai Martyrdom pays homage to the Revolutionary Martyrs of the New Democracy with a strong red memorial atmosphere and includes the park's northern gate which highlights this solemnity. The park is also a famous scenic tourist area in Nanjing and needs to reflect its publicity and openness.

立面图 Elevation

大门正立面实景 Front view of the gate

The building took on an extremely modern and minimalist form, with four volumes set in a 150-meter-wide urban interface – two ticket booths, a duty room and a canteen. The four building volumes house different functions and divide the south entrance into three sections horizontally. The buildings and fences appear alternately to show an intended spatial-division along with an effort to promote integration within its context while also upholding closeness and rituality. Based on simple shapes, Xiang Bingren carefully considered materials and details. Stone and glass were selected to reinforce the contrast between the solid and void while also amplifying the contrast between the historical and modern. The hollowed partition of granite facades adopts different horizontal widths, lending a sense of delicacy and calmness to the overall solemnness. The two ends of the gate are freely inlaid with many circular stone-fiber optic lights, providing a visual demonstration to the line "petals fall like rain, stones get speckled by raindrops", fully demonstrating the innovation and contemporary nature of Nanjing. With a simple and powerful volume composition, open and flexible space interface, and distinctive facade details, the building itself seems to be growing out of Yuhuatai naturally while complimenting the city perfectly.

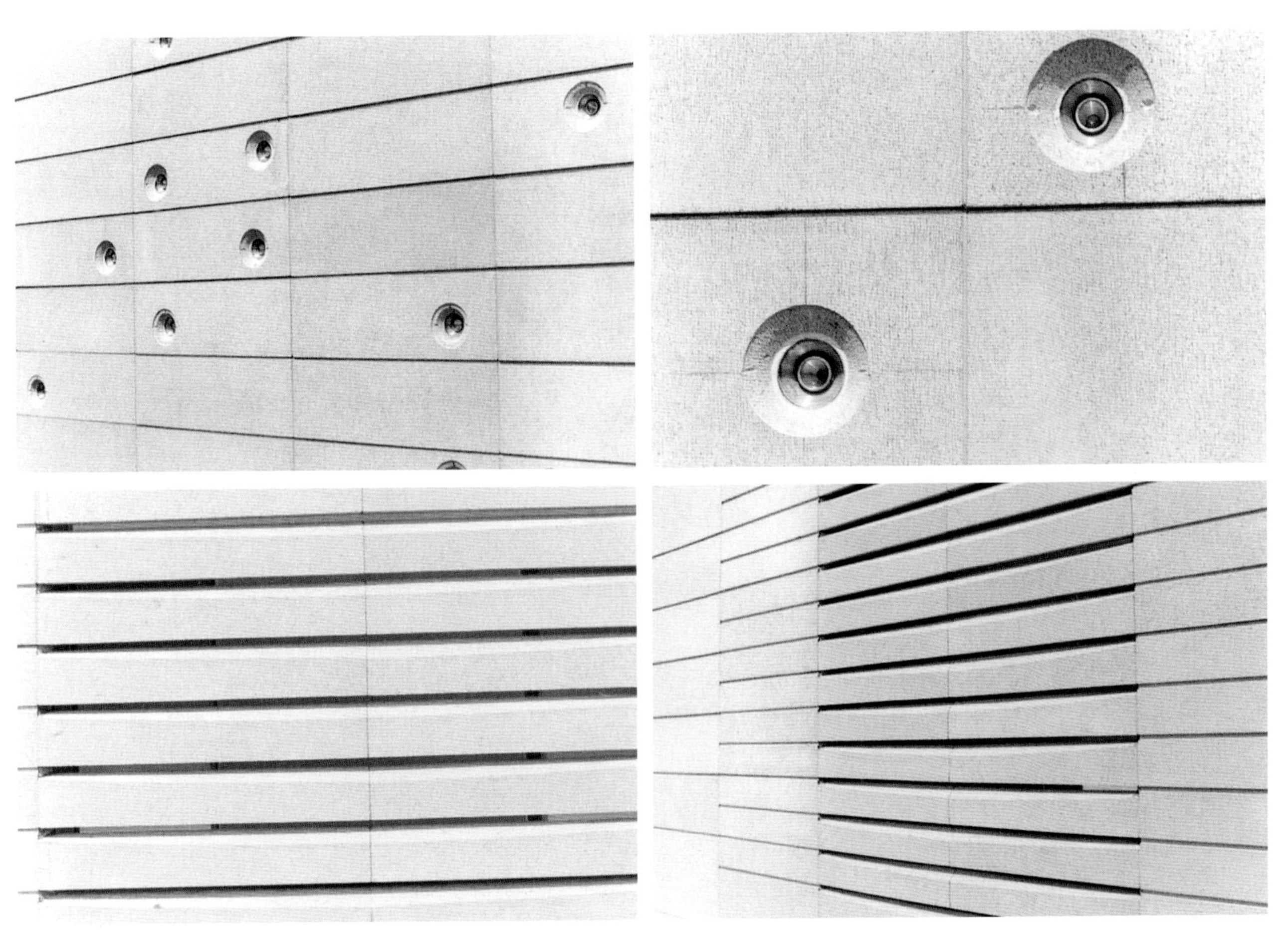

立面细节 Details

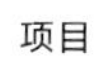

^ 售票厅 Front gate details
∨ 入口夜景 Night view

# 评论
Review

舒展的墙体延伸至两侧的自然土坡，与大地相融，强调出平缓感，这种感觉由于墙面石材之间的水平凹缝得到了进一步加强。洗练的语言以一种中性、对称的姿态营造出肃穆的氛围。厚实的墙面夹着轻盈的玻璃体，虚实对比消解了沉闷感。与北大门的古典气质不同，南大门更为现代，但在空间上又与整个陵园十分契合。

陈强
同济大学建筑城规学院硕士生导师
上海道辰建筑师事务所（DCA）主持建筑师

南京雨花台南大门的设计使城市配套服务功能与公共空间体验质感结合在一起，以严谨的功能形体构成结合灵动的细部设计（自由分布的圆形内凹光纤灯位和倾斜的石材镂空切口）构建出恰如其分的情感氛围（纪念性与公共性）。在本世纪初中国城市建设竭尽所能关注于“从无到有”之时，项秉仁先生却在公园大门这样的微设计中义无反顾地执着于“从有到精”的创造，实在难能可贵。而其严苛但饱含热情的空间与细部刻画，在当前逐渐回归现代主义基本追求的世俗建筑设计需求趋势下，更突显出前瞻性。

郭红旗
上海一砼建筑规划设计有限公司创始合伙人、设计总监
建筑学博士

The stretched wall extends to the natural slopes on both sides, blending with the earth and emphasizing a sense of calmness, which is further enhanced by the horizontal seams of its stone surface. The refined design contributes to a solemn atmosphere in a neutral, symmetrical posture. The light glass body between the thick walls creates a contrast between the void and solid, dispelling dullness. Unlike the classical temperament of the North Gate, the South Gate is modern yet spatially in harmony with the entire cemetery.

Chen Qiang
Master Supervisor, CAUP, Tongji University
Chief Architect, DCA Studio

Though encountered regularly in our daily lives, park gates seem just as frequently overlooked. The design of the south entrance of Yuhuatai Park Gate combines the infrastructure nature of the gate with better spatial experience to properly construct a sentimental atmosphere that is monumental yet can be publicly enjoyed, which is achieved by rigorous volume composition and dynamic detail (freely distributed circular concave fiber optic lamp position and inclined stone hollow incision). While China was focused on the rapid development of urban spaces during the beginning of the century, Xiang should be commended in his rare efforts to achieve better quality for a micro-design such as a park gate. His rigorous yet enthusiastic attitude for space and details seems extremely forward-looking against the backdrop of secular architectural design's gradual return to the basic pursuit of modernism.

Guo Hongqi
Founding Partner/ Design Supervisor, Shanghai Yitong Design Co., Ltd.
Doctor of Architecture

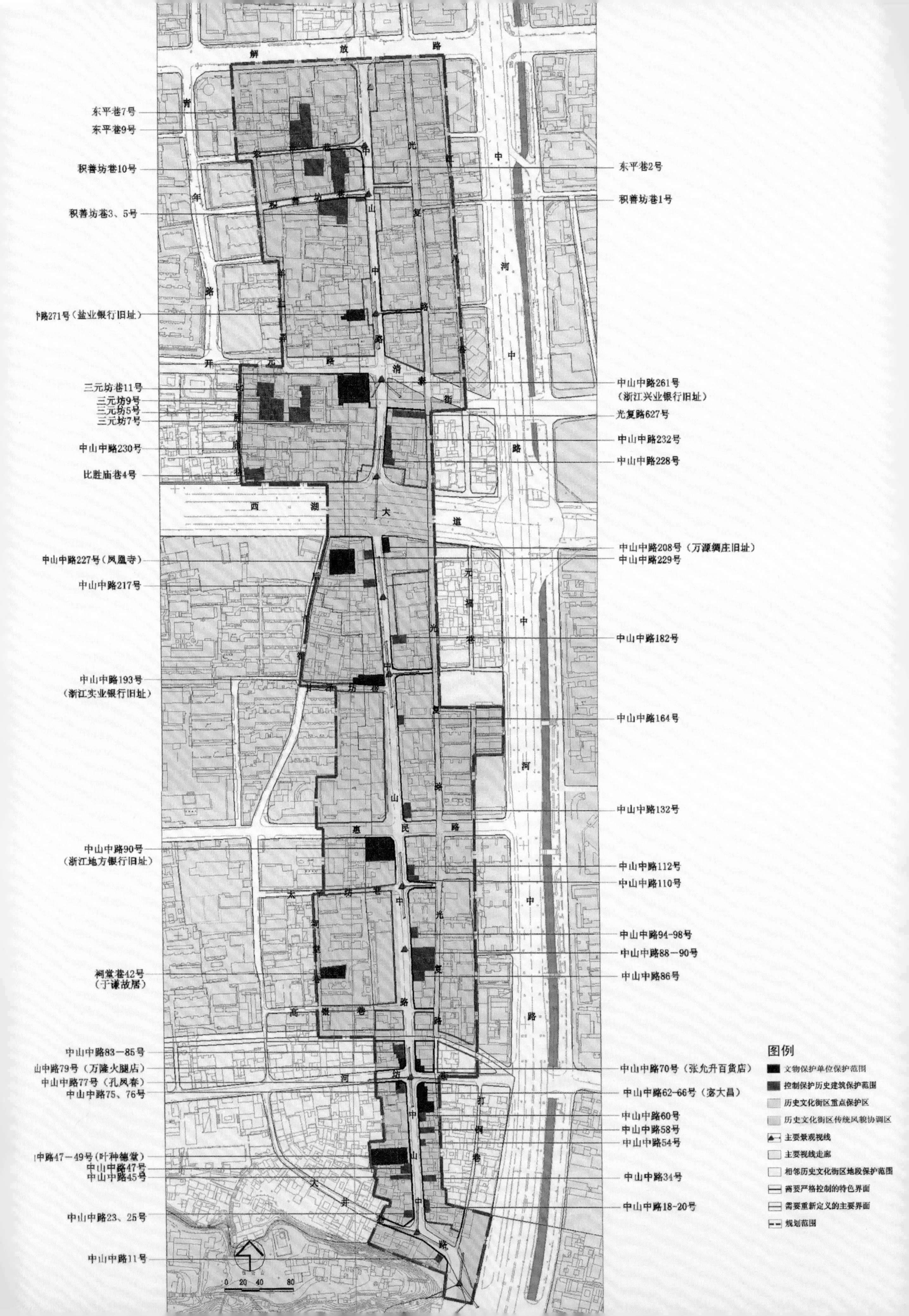

解放路
西湖大道
中河中路
光复路
中山中路
东平巷7号
东平巷9号
积善坊巷10号
积善坊巷3、5号
路271号（盐业银行旧址）
三元坊巷11号
三元坊9号
三元坊5号
三元坊7号
中山中路230号
比胜庙巷4号
中山中路227号（凤凰寺）
中山中路217号
中山中路193号
（浙江实业银行旧址）
中山中路90号
（浙江地方银行旧址）
祠堂巷42号
（于谦故居）
中山中路83－85号
山中路79号（万隆火腿店）
中山中路77号（孔凤春）
中山中路75、76号
中路47－49号（叶种德堂）
中山中路47号
中山中路45号
中山中路23、25号
中山中路11号
东平巷2号
积善坊巷1号
中山中路261号
（浙江兴业银行旧址）
光复路627号
中山中路232号
中山中路228号
中山中路208号（万源绸庄旧址）
中山中路229号
中山中路182号
中山中路164号
中山中路132号
中山中路112号
中山中路110号
中山中路94-98号
中山中路88-90号
中山中路86号
中山中路70号（张允升百货店）
中山中路62-66号（宓大昌）
中山中路60号
中山中路58号
中山中路54号
中山中路34号
中山中路18-20号
0 20 40 80
图例
文物保护单位保护范围
控制保护历史建筑保护范围
历史文化街区重点保护区
历史文化街区传统风貌协调区
主要景观视线
主要视线走廊
相邻历史文化街区地段保护范围
需要严格控制的特色界面
需要重新定义的主要界面
规划范围

# Protection and Renovation for the Historic Neighborhood of Mid. Zhongshan Rd., Hangzhou



2006 / 2007

# 杭州中山中路历史街区保护与更新

2006年，项秉仁接受杭州市规划局的委托，主持杭州中山中路历史街区的保护和更新设计。中山中路历史街区位于吴山北麓，历史上一直是杭州城区的中心和商贾云集之地。1949年以后，随着杭州近代商业中心的北移和西迁，中山中路逐渐失去了城市商业中心的地位，20世纪末的城市改造更是加速了这一进程，功能衰退，风貌破败。

中山中路历史街区城市设计是在相关保护与整治规划策略的指导下，对中山中路保护整治成果在形态上进行完善和深化，其目的在于从建筑与城市空间形态的角度出发，科学有效地落实相关保护规划的具体要求，其成果是对历史街区形态的总体控制，并对今后中山中路历史街区整治实践起到指导作用。

成功完成这一项目的前提是建立一套分析和解决问题的思路。首先便是要明晰城市设计范围内街区的历史资源和价值，在调查街区城市空间与建筑现状的基础上建立城市设计的目标和理念，在特色街区的定位、形态、用地、交通等方面提出具体的城市设计策略。然后采取分段设计的方法，将中山路分为三个空间段落和八大特色主题街区，并进行用地规划调整、交通规划调整、空间形态控制，最后通过点线面的深化设计及针对街道空间景观要素提出具体的设计导则。

In 2006, Xiang Bingren accepted the commission from the Hangzhou Municipal Planning Bureau and presided over the protection and renovation of the Hangzhou Mid. Zhongshan Road Historic Neighborhood. Mid. Zhongshan Road Historic District at located in the northern foot of Wu Hill. This block, historically the center of Hangzhou with dense commercial activities, was once very prosperous. After the founding of the People's Republic of China, Mid. Zhongshan Road's role as the commercial center gradually faded as trading activities moved northward and westward. The urban transformation during the late 20th century accelerated the process, causing the functional decline and dilapidation of this area.

# 杭州中山中路历史街区保护与更新

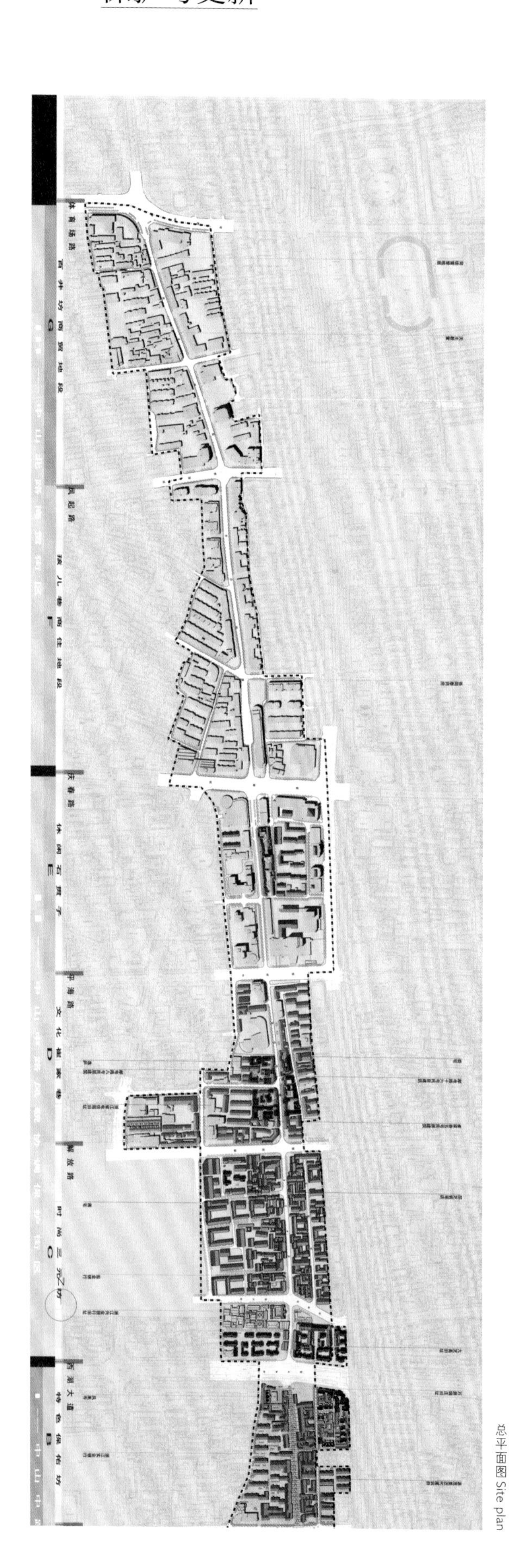

总平面图 Site plan

^ 沿街长立面东 East elevation
∨ 沿街长立面西 West elevation

# Protection and Renovation for the Historic Neighborhood of Mid. Zhongshan Rd., Hangzhou

Guided by protective planning documents, Mid. Zhongshan Road Historic Neighborhood was designed to complete and continue the current advances in protection and renovation of its urban surroundings. It was further meant to scientifically and effectively implement specific regulatory requirements from the perspective of architecture and urban spatial form. The urban design process should result in a guideline posing an overall control over the pattern of the historic district and regulating future practices for the renovation of the historic neighborhood of Mid. Zhongshan Road.

The premise of successfully completing this project was to establish a set of ideas for analyzing and solving a myriad of problems. The first step was to clarify the historical resources and values of the relevant neighborhoods. Based on a survey of the surrounding urban space and building status, the design team established the goals and concepts of urban design by putting forward specific urban design strategies including the positioning, shape, site and traffic of special blocks. Mid. Zhongshan Road was then divided into three sections, eight special themed blocks. Zoning adjustments were also carried out in addition to traffic planning adjustments and spatial form control. Finally, through much more detailed design of points, lines and surfaces, specific spatial and landscape design guidelines for the street were gradually developed and implemented.

模型鸟瞰 Bird's-eye perspective of model

# Protection and Renovation for the Historic Neighborhood of Mid. Zhongshan Rd., Hangzhou



## 评论
## Review

南北横贯杭州主城区的中山中路，这条曾专供皇帝行走的“御街”，可谓饱经历朝历代风雨，浸润千百年来市声。多年来，在杭州市各级政府领导、专家、设计师的共同努力之下，强化“以古为魂”的理念、发挥“由古为名”的优势、做好“借古而兴”的文章、打造“因古而靓”的历史街区新风貌。2006年的夏天，项老师团队承担起中山中路历史街区保护与更新的项目并着手现场调查和测绘起始，设计思路和成果经过多次专家评审和修改，历经约一年半的时间，最终得到了杭州市政府领导和有关专家的认可及相关主管部门的批准，并作为下一阶段保护和更新工作的指导文件。保护和更新后的中山中路将会是一条集中体现杭州古城特色，反映杭州民国时期沿街商业建筑风貌，具有杭州独特文化和历史的传统综合商住街区。同时，它也将会成为一条能延续城市历史文脉、缝合新旧街区间的裂缝、集中反映杭州历史变迁和体现风貌价值的街区之一。

祁涛

深圳华森设计执行总建筑师
建筑学博士

As a major north-south thoroughfare crossing the heart of Hangzhou, Mid. Zhongshan Road was designed as the "Royal Street" for the emperor to walk on. It has played witness to the important history of the city for over a thousand years. During this time Hangzhou government officials, experts, and designers worked together to implement the philosophy of “history as soul”. We strove to take advantage of Hangzhou’s “historical reputation”, to carry out the plan of “learning from history for a prosperous future ", and also to create a new look for historic neighborhoods that “derive beauty from the past”. Since the summer of 2006, Prof. Xiang's design team has taken on the project of protecting and renovating Mid. Zhongshan Road’s Historic Neighborhood. The team additionally conducted on-site investigations while surveying, mapping, and adjusting the design concept and scheme on many occasions under the advice of experts. After a year and a half of hard work, the design scheme finally won the approval from the head of the Hangzhou municipal government, experts, and authorities, and now serves as the guidance document for the next stage of protection and renovation projects. The protected and regenerated Mid. Zhongshan Road serves as a traditional and comprehensive commercial and residential block that embodies the unique features of the ancient city of Hangzhou, and reflects the commercial architecture along the streets of Hangzhou during the Republican era along with its unique culture and history. At the same time, the project will also become one of the most representative neighborhoods that can reflect the historical changes and the style of Hangzhou.

Qi Tao

Executive Chief Architect , HS Architects Shenzhen
Doctor of Architecture

鸟瞰图 Aerial view

# Conservation and Renovation for the Yuanfuxiang Historic Block, Hangzhou

2006 /

# 杭州元福巷历史街区保护更新

几乎是与杭州中山中路历史街区保护与更新项目同一时期，项秉仁也接受了杭州元福巷历史街区保护更新计划的委托。项目基地位于西湖大道和中河路交叉口，距西湖约900米，由光复路、中河路、清泰街、保佑桥东弄四条街巷围合而成。街区内部现存有一些晚清和民国以来的历史建筑，形式多样，有传统浙江民居，也有西式石库门和风格上中西合璧的商业建筑。2004年杭州市规划局将元福巷为首的13个街区正式列为历史街区，而该项目的主持建筑师的工作任务，是在政策法规和已有的详细规划的控制下展开更具体的三度空间和形体的城市设计及建筑设计。换句话说，也就是承担了从理论政策条文和字面控制向真实具象的物质形态转换的关键工作。建筑师的水准和建筑设计的品质对于历史街区改造和更新的成败至关重要。

在总体规划和建筑设计中，项秉仁特别注重历史街区城市空间形态与建筑形态的处理，避免在设计历史街区中更新或新建的建筑物时一味强调将建筑形式恢复到某一特定时期，使之成为某个时代的仿古街，最终导致真假不分，使历史街区的发展连续性消失殆尽。保护并不意味着简单的复制和延续，历史街区必须在原有格局和新格局之间做出妥协和取舍，这种新的格局生成更像是一种拼贴的过程，新的建筑在老的街区格局的空隙范围内，与其相互交织并最终形成一个完整的历史街区更新格局。

对于在历史街区内引入与历史建筑对比鲜明的新建筑形式应是谨慎和理性的。在元福巷地块的改造中新建建筑无论在形态还是在空间处理上都是以延续协调作为设计态度，以将传统建筑形式进行现代转译作为设计原则。建筑形式源于对基地内传统民居的研究，将传统木构作为基本建筑尺度所控制的“间”的做法在新建建筑中加以延续；新建建筑外观强调构成元素（黛瓦、粉墙、原色小木作三元素）权重比例的重构，从而使其摆脱了原有的形式制约而产生非传统的构成方式。

沿清泰街局部立面图 Elevation along Qingtai road

沿清泰街透视图 View of main entrance

此外，新建筑在细部设计上也注重原真性和创新性。在比例尺度构成方式上与老建筑相协调，但在具体的节点构造材料上采用新的做法，以期达到在历史街区中历史保护建筑和新建筑在形态上既协调又不混淆的要求。保留历史街区街巷空间的原有格局及风格，通过对门、廊、窗、墙面、屋顶等建筑构件和建筑界面的设计丰富地块传统的空间形式，增强历史街区的氛围和认同感。建筑单体设计吸收保留历史建筑的特征要素，在更新建筑设计中加以继承、体现，使保留和更新建筑总体风貌和谐，形成完整的历史街区人文景观。

At almost the same time Xiang Bingren was undertaking the protection and renovation of the Mid. Zhongshan Road historic district, he also accepted the commission for the conservation and renovation of the Yuanfuxiang Historic Block in Hangzhou. The site is located at the intersection of Xihu Avenue and Zhonghe Road, about 900 meters away from the West Lake. It is surrounded by four roads: Guangfu Road, Zhonghe Road, Qingtai Street and East Baoyouqiao Lane. There are some historic buildings in the block from the late Qing Dynasty and the Republic of China, including traditional Zhejiang dwellings, Shikumen buildings, and commercial buildings bearing both Chinese and Western characteristics. In 2004, the Planning Bureau of Hangzhou officially listed Yuanfuxiang among 12

# Conservation and Renovation for the Yuanfuxiang Historic Block, Hangzhou

沿光复西路透视 View of Guangfu west road

other blocks as historic blocks. As leading architect on the project, Xiang's main task was to develop a three-dimensional urban design along with architectural products according to the current regulations and detailed planning documents. In other words, Mr. Xiang took on the key role of interpreting written regulatory policies into tangible physical forms. Both the capability of the architect and the quality of the design were directly associated with the success of this conservation and renovation project.

During the master planning and architectural design process, Mr. Xiang paid special attention to the design of urban and architectural forms in such a historic block so as to avoid the mere mimicking of architectural styles of a certain time when designing new buildings. He worked to blur the difference between fake and authentic, while eliminating the intrinsic consistency of the historic blocks' development. Conservation was interpreted as meaning more than a simple reproduction and continuation. The historic block needed to choose between the original pattern and the new pattern, generating a collage of various elements. The new buildings are integrated with the existing old city fabric and eventually helped to form a new and complete pattern for these historic blocks.

Architects should be very cautious and rational when introducing new architectural forms that contrast with historic buildings in historic districts. In the transformation of the Yuanfuxiang block, the design principles for new construction are: continuation and coordination with

沿元福巷透视图 View of Yuanfuxiang

the historic fabric, modern interpretation of the traditional architectural form. The architectural form is derived from the study of traditional dwelling on the site, which continues the practice of traditional wooden structures controlled by module construction standards. While the appearance of new buildings emphasizes three elements of traditional architecture, black tiles, white walls and unpainted small wooden work, it interprets them differently, so that new buildings can shrug off of the constraints of traditional norms, resulting in novel compositions.

In addition, detailed designs of new buildings also focus on authenticity and innovation. The new building is coordinated with the old one in terms of volume and proportion, but new construction materials were selected in order to achieve a morphologically balanced and clear coexistence of historic buildings and new buildings. The design of the street space retains the original pattern and style of the historic block. At the same time, the designer enriches the traditional spatial form of the site through the new design of building components and architectural interfaces such as doors, corridors, windows, walls and roofs. These additions and elements enhance the atmosphere and emphasize the unique identity of the historic block. The design of the building absorbs and retains the characteristic elements of the historic building to achieve an overall harmonious style and to form a complete humanistic landscape of the historic block.

# Conservation and Renovation for the Yuanfuxiang Historic Block, Hangzhou



## 评论
## Review

谨慎的地域性建筑的现代转译是元福巷改造项目设计上的独到之处，在改造更新非常保守的时期，这个项目积极地保留了原有历史街区的空间肌理，在新建和保留建筑之间，传统建筑的建构逻辑成为了新建建筑的形式构成出发点，新的建筑材料和工艺以及功能带来的违和感在建筑师老到的处理下成为新旧建筑新旧有别、各自精彩的积极元素。

吴欣

天华集团副总建筑师
上海天华执行总建筑师
建筑学博士

The Yuanfuxiang renovation project is exceptional because of its cautious yet regional interpretation of modern architecture. During a period of time when renovation was fairly conservative, this project actively retained the spatial texture of the original historical block while also finding the formal logic of new buildings in traditional architectonic treatments. The seeming incompatibility of new building materials and techniques has been transformed quite skillfully into exciting elements indicating the difference between old and new.

Wu Xin

Vice Chief Architect, Tianhua Group
Executive Chief Architect, Shanghai Tianhua
Doctor of Architecture

# 3

## 人居环境建筑：传统内涵和诗意

进入新千年后，项秉仁的建筑设计事业慢慢步入成熟期，虽然初始阶段的诸多作品并不具有统一的风格，但都在回应各自所处的社会场景与文化语境，也开始具有独立且完善的形式语言系统。在人居模式和生活理念的探索之路上，杭州富春山居（2000）是对浙江民居的传统语汇的吸收和再创造；广州大一山庄（2006）是现代集群设计的一次实践；千岛湖润和建国度假酒店（2004）是对建筑与自然“和谐生长”的挑战；佘山珑原别墅（2010）是对现代空间中式意境的初探，而旺山六境（2011）则是一场关于居住建筑类型学创新的思辨。从中不难看出项秉仁在不断地回应“传统与当下，全球化与本土化”等话题，这种源于现代主义精神的主体性创作思维，在对于地形气候、景观环境、地域文化、时代特色及室内场景的全方位渗透和把握下，更好地实现了项秉仁对此类建筑的人文情怀的追求。

## Architecture for Habitation: Traditional Implication and Poetics

Xiang's architectural design career began to mature when it enters the new millennium. Although many of his works in the initial stage did not have a unified style, they all responded to their respective social and cultural contexts, and also began to develop an independent and complete formal language system. On the way of exploring habitation models and life concepts, Fuchun Mountain Villa, Hangzhou (2000) is absorption and re-creation of the traditional residential vocabulary in Zhejiang; Dayi Mountain Villa, Guangzhou (2006) is an attempt to design a modern complex; Qiandaohu Runhe Jianguo Hotel (2004) is a fruit of “harmonious growth” of architecture and nature; Sheshan Longyuan Villa (2010) is an integration of Chinese artistic conception into modern spaces; and finally, The Six Villas, Wangshan(2011) is a speculation on the typological innovation of residential architecture. Xiang constantly addresses topics such as tradition and present, globalization and localization in his projects. Thanks to this subjective way of creative thinking that originates from the spirit of modernism, Xiang has fulfilled his humanistic pursuits in buildings through comprehensive understanding of terrains, climates, landscape environments, regional cultures, characteristics of the times and indoor scenes.

建筑与山地的自然融合 A villa sitting well on its hill site below

# Fuchun Mountain Villa, Hangzhou



2000 / 2004

# 杭州富春山居别墅区

2000年杭州富春山居别墅区的设计是项秉仁回国后做的一次将中西方建筑理念相融合的尝试。项目基地位于富阳银湖开发区，毗邻杭州市区的一处丘陵缓坡地带，占地1500亩，植被良好，水系丰富，计划修建独立别墅、联排别墅900余套。当时的杭州房地产业发展快速，城市建设已在国内率先迈入了住宅城郊化的探索，这个项目也成为摆在项秉仁面前的一个新课题。西方欧美国家在经历了多年无节制的郊区住宅开发之后，已经体会到了这种无序开发所带来的弊病。为了避免诸如"卧城""死城"的失败案例再次出现，项秉仁希望吸收新城市主义的规划理论，在项目规划上确定合理的开发规模，构建完善的配套公建和交通体系，塑造舒适的居住氛围。

总体规划根据环境、景观、朝向、植被、风水等资源条件将用地划分为若干级别，规划相匹配的不同面积区间和档次的物业类型，并由此确定不同物业的开发数量和比例。在此基础上，采用了在国外规划中行之有效的组团式规划模式，进行小区空间结构和道路系统的规划。它一方面保障了住户的安静、安全和私密，最大限度地避免了主要道路对住户前区的穿越。另一方面使得组团间可以有大量的绿化隔离空间，优化了每户别墅的周围环境。规划设计在对现有的自然景观特色加以保护和强化的同时，特别强调小区内的公建和景观，中央大面积集中绿地和湖面的共享，为住区增添生活气息和凝聚力。

别墅单体的功能配置及空间设计紧密结合环境地形，最大程度地适应现代中国人的生活方式。在建筑设计中注重各功能用房的空间需求和相互关系，着意营造室内空间的艺术性和变化有序的空间序列。上空的玄关、挑空的起居室和落地的大玻璃窗使得上下内外的空间融为一体，实现"依山傍水，保留原木，把家轻轻放入大自然"的设计初衷。

富春山居的另一大亮点就是造型设计的与众不同。没有盲目追随当时杭州城内高尚地段的西式豪宅别墅区风貌，而是力图探索一种与时代、地区相适应的建筑风格，沿承传统浙江民居的建筑文化遗产，利用坡屋顶的空间和外形特点因地制宜地构建单体建筑形态，创造具有浙江地方风情的现代住宅。批判和创新的精神，颇有当年赖特嘲讽仿效欧洲贵族气派、追随维多利亚传统，极力创造美国本土建筑文化时的风范。

---

The design of the Fuchun Mountain Villa in 2000 was Xiang Bingren's first attempt at combining Western with Chinese architectural ideals since his return from America. The site is located in the Fuyang Yinhu Development Zone, a piece of hilly land around 1,000,000m² with

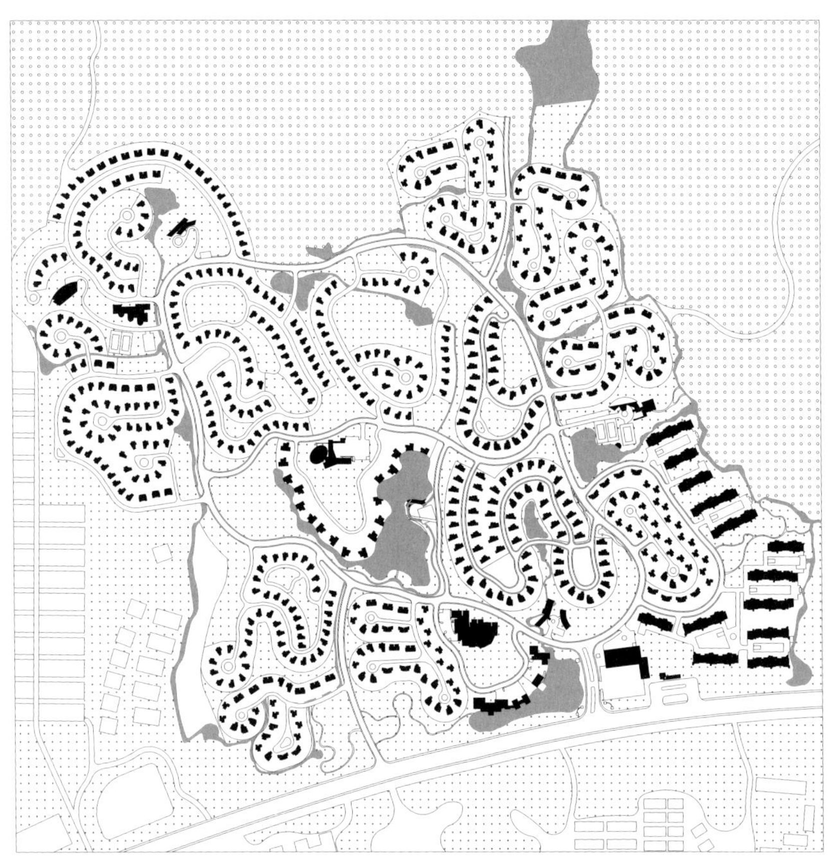


总平面图 Site plan

good vegetation and several bodies of water also adjacent to downtown Hangzhou. According to the client's plan, around 900 units of single family houses and townhouses as well as ancillary facilities would be built. At that time, real estate development in Hangzhou was booming as the city planners began to explore suburbanized housing, a relatively new topic for Xiang Bingren. Western countries had already learned their lessons from many failed expansions into suburban areas and Xiang Bingren wanted to prevent "sleeping" and "dead" cities from happening again. To this end, he studied new urbanism theories in planning and worked out a reasonable scale for development and impeccable supporting facilities as well as a transportation system to create a comfortable living atmosphere.

Out of consideration for the environment, landscape, orientation, vegetation and Feng Shui, the land was divided into different pieces with different grades which supported different housing types with different floor plans and prices accordingly. Based on this subdivision, Mr. Xiang adopted the time-tested cluster planning strategy from the West. On the one hand, this strategy could guarantee the tranquility, security and privacy of residents by minimizing the exposure of housing to main streets. On the other hand, the strategy allows more space for greenery between housing clusters in order to improve the environment around the houses. The master plan not only preserves and optimizes the natural scenery of the site but also emphasizes public building and public accessible landscapes, such as the shared central park and lake, which can catalyze liveliness and cohesion of the neighborhood.

On an architectural level, the spatial and programmatic designs are highly associated with topographic and environmental characteristics to fulfill the needs of a contemporary Chinese lifestyle. The intrinsic spatial needs of different functions and their mutual support were taken into consideration, and an artisanal yet organized spatial sequence was created carefully in response. For instance, the double-height design at the living room and foyer combined with roof-to-floor glasses successfully blur the boundaries between inside and outside, upstairs and downstairs, echoing the initial intension of residing among nature with less disruption.

One of the selling points of the Fuchun Mountain Villa is its distinctive appearance. At the time, a Western mansion style was prevalent in the prime real estate development of Hangzhou. Despite its popularity, Xiang Bingren chose not to follow the prevailing mode, opting instead to explore a more apropos architectural style that could nest carefully with the time while remaining relevant to the location. Xiang inherited the legacy of traditional Zhejiang residential buildings by taking on the signature shed roof and generating form according to local conditions which led to a modern housing design that upholds a vernacular character. The critical innovative spirit reflected in this design echoes Wright's efforts to create an original American architectural culture in defiance of the prevalence of simple imitations of Victorian aristocratic styles.

# Fuchun Mountain Villa, Hangzhou

实景照片 Photos

手绘草图 Freehand sketch

实景照片 Photos

手绘草图 Freehand sketch

## 评论
## Review

随着中国城市化进程的加速，国内房地产业逐步进入黄金发展期，当梦想西方生活方式的新兴阶层还迷恋于当时泛滥的欧式别墅时，富春山居犹如一股清风。作为国内最早研究赖特的学者，项先生或许在尝试一种比拟于赖特式“美国风”（Usonian）的别墅类型，加上其境外生活工作十年的经历，使得建筑试图呈现出美式文化与当地传统民居的结合。具有乡野气息的毛面石材与浅色墙面的肌理对比、巨大挑台与宽缓坡顶对起伏地形的回应，都呈现了对自然的思考和对地域性的当代诠释。

陈强

同济大学建筑城规学院硕士生导师
上海道辰建筑师事务所（DCA）主持建筑师

With the accelerating pace of urbanization in China, the domestic real estate industry has entered a golden age. The Fuchun Mountain Villa was like a refreshing breeze, a point of difference using a time when the emerging middle class was still obsessed with a Western lifestyle and European villas. As the first scholar to have discussed his studies on Wright in China, Mr. Xiang was trying to design a new type of villa that is similar to the Wright-style “Usonian”, plus his 10-year experience of living abroad, leading to a design that resembled American culture and local traditional dwellings. The contrast between the texture of the rough-faced stone and the light-colored wall, the huge platform and the gently sloping top all respond to the undulating terrain, presenting a reflection on nature and a contemporary interpretation of regionality.

Chen Qiang

Master Supervisor, CAUP, Tongji University
Chief Architect, DCA Studio

S11 透视效果图 Perspective rendering of S11

# Dayi Mountain Villa, Guangzhou



2006 / 2009

# 广州大一山庄

大一山庄位于广州市白云山麓，是著名风景区内的别墅地产开发项目。其中一期约20栋独栋别墅由国内外知名设计师担纲设计，项秉仁参与设计了两栋别墅，并最终建成了其中的一栋。这一作品是其投身集群设计（21世纪初一度风靡中国的以国内外先锋建筑师为主体的集体设计实践）的一次重要实践。这个系列作品虽然体量不大，但显示了他一贯的设计追求。

项目整体规划为山地别墅居住区，项秉仁设计了两栋约800平方米的独栋别墅，都是地上两层、地下一层的简洁正交体量。其中一栋为长方体，所有功能在内部有机组织，外部仅通过表皮的虚实与景观界面发生关联。另一栋为L型，其两翼亦为完整的长方体，没有设置任何退台等可能改变方正体量的元素，仅设置了一些悬挑的阳台及由竖向格栅围合的灰空间。两栋建筑不约而同的高度理性而内敛，类似于萨伏伊别墅与外部景观的关系，而绝非塔里埃森的场所设计方式。这种对外部环境的应对方式让我们看到纯正现代主义甚至功能主义的烙印，虽然试图用表皮去化解体量的孤立，但总体而言还是高度克制和内省的建筑语言，这同许多房地产开发中的别墅设计手法大相径庭，体现出一种实验建筑师的倾向，以自我为中心的建筑语言的表述，也可以看成是对安藤忠雄、博塔等同辈大师的共鸣。

有趣的是与内敛的形体相反，L形别墅的表皮设计语言完全采取了不同的态度，甚至可以看到后现代美学的影子——红砖、浅色石材、木格栅、深色工字钢、落地玻璃幕墙拼贴和叠合在一起，形成了材料和视觉上的生动性。虽然遗憾的是它并未建成，但项秉仁在之后实践中的很多设计倾向已在此展现出来——对大片玻璃幕墙系统的偏好（宁波市东部新城行政中心），对于多种材料在同一个项目中并置的倾向（宁波文化广场），等等。

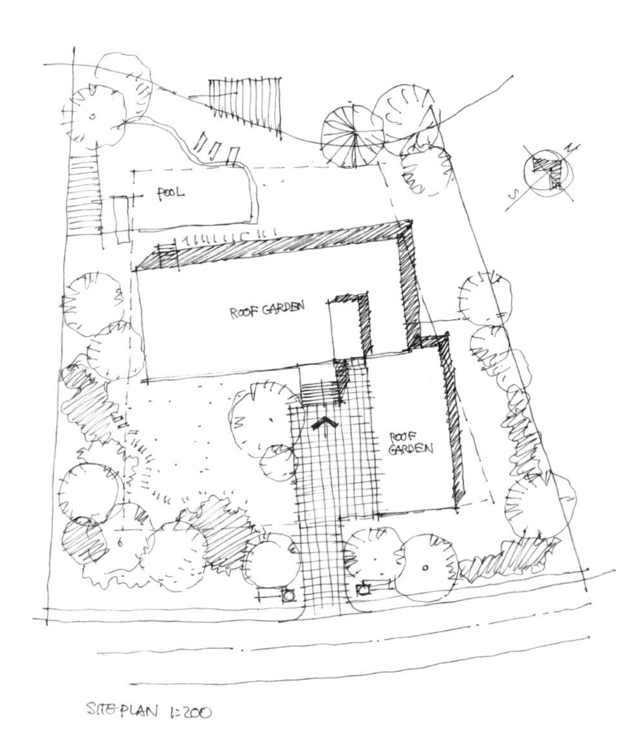


S11 手绘总平面 Site plan

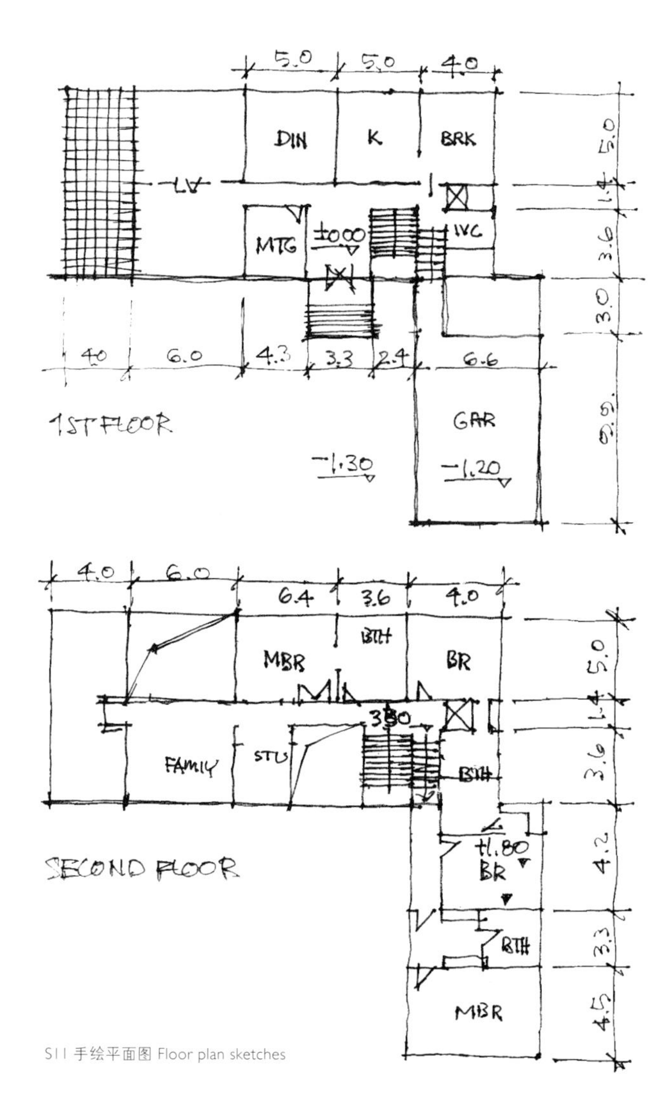


S11 手绘平面图 Floor plan sketches

Dayi Mountain Villa, a real estate development project located at the famous scenic Baiyun Mountain, Guangzhou. The first stage of the project, including about 20 villas, was designed by well-known designers and design institutes both domestic and foreign. This project is an important project in Xiang's architectural group design, which was very popular in China in the early 21st century and widely attended by avant-garde architects worldwide. Though the project was small, it suggests Xiang's later career path.

The whole development was planned as a residential area featuring mountain villas. Xiang Bingren designed two individual villas both with floor areas of approximately 800m² over 3 stories, two above-ground and one under. One of them is a cuboid box with all the functionalities organized internally and only interacts with the landscape through the voids on its facade. The other villa is L-shaped composed of two perfect cuboid wings with few balconies and grate defined grey areas. The designs of both villas are proportional and introverted, echoing the way Villa Savoye interacts with its landscape, but different from Taliesin's original ideals. This attitude towards the context recalls modernist and functionalist preoccupations. Though the facade design attempts to reconcile with its surroundings, the overall design strategy is still highly restrained and introverted, different from the villas designed by most real estate developers. In a way, Xiang's design is vaguely experimental, featuring a personal

# Dayi Mountain Villa, Guangzhou

S1 透视效果图 Exterior view

expression of architectural language while resonating with works of his coetaneous masters, such as Tadao Ando, Mario Botta.

Interestingly, contrary to its introverted volume, the facade of the L-shaped villa adopts a different design attitude, showing traces of a post-modernist aesthetic: red bricks, light-colored stone, wooden grating, dark-colored joist steel, and a glass curtain wall are combined and overlapped, presenting a visual and material vividness. Although the design was never realized, it reveals many of Xiang's design preferences: the extensive use of a glass curtain wall system (as in the Ningbo New Municipal Center Design), and the juxtaposition of various materials in one project (as in the Ningbo Cultural Plaza).

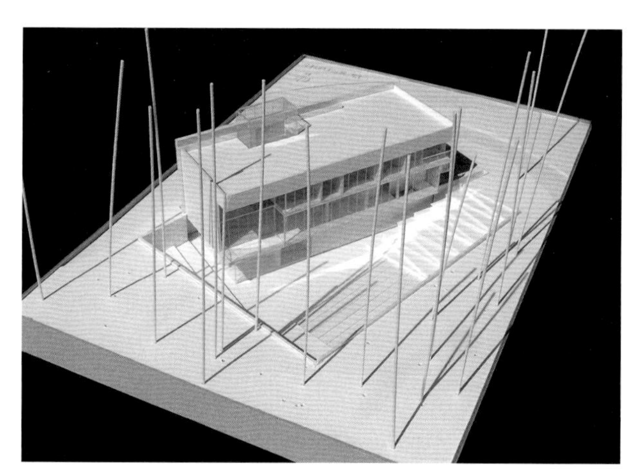

S1 过程模型图 Working model

< 餐厅空间 Living room
> 客厅空间 Dining room

# 评论
# Review

项老师的大一山庄显得内敛、平静，且有力量。建筑采用经典的现代主义设计手法，形体方整简练。两座别墅入户均抬高1.2米左右，既能有效缓解南方地区潮湿的气候环境问题，改善地下空间的采光，也使得主体建筑与自然环境之间达成一种交互式的融合，颇有点向现代主义建筑大师密斯致敬的意味，可谓一举多得。别墅室内采用流动的空间设计手法，入户玄关、楼梯、餐客厅、厨房等功能巧妙地融于一体，在平面及三维立体空间中，视线起承转折，流而不通、隔而不断，空间体验极为丰富。

大一山庄通过几何形体的穿插冲突、立面虚实的对比变化、建筑新材料的大胆运用、细部比例的考究处理，既体现了老一辈建筑师深厚的设计功力与专业素养，同时又不乏大胆创新。

肖志抡

上海仑城建筑规划设计事务所主持建筑师

Restrained, calm yet powerful. This is what one feels in Xiang's Dayi Villa, a classic modernist building with a neat and concise shape. Both volumes are raised by about 1.2m at the entrance. This action proves to serve multiple purposes: it effectively alleviates the humidity in the south region and improves the lighting conditions in underground spaces, and it enables an interaction between the main building and the surroundings, almost like a distant nod to the modernist pioneer Mies. The interior spaces are designed to be free-flowing. Various functions such as entrance hallway, stairs, dining/living room, kitchen, etc. are cleverly integrated into one, therefore providing an extremely rich spatial experience for visitors along the route, with views flowing but not mixed, separated but not isolated.

A wealth of design skills and professionalism can be seen here through the interspersed conflict of geometries, the contrasting changes of virtual and real in the facades, the daring use of new building materials, and the meticulous treatment of details and proportions. They might as well demonstrate the constant innovation of an older generation of architects in this rapidly changing era.

Xiao Zhilun

Chief Architect, LAD Partner

因地制宜与自然融合的建筑形体 Hotel building well fitting its site to have lake view for most guestrooms

# Qiandaohu Runhe Jianguo Hotel

2004 / 2012

# 千岛湖润和建国度假酒店

千岛湖润和建国度假酒店位于千岛湖风景区东西向的半岛——梦姑岛，可以说是项秉仁遇到的地形地貌最为复杂的一个项目。狭长的梦姑半岛伸入开阔的千岛湖中心湖区，岛屿南、西、北三面环湖，东面则接临直通千岛湖镇的城市道路。岛上风景秀丽，植被丰茂，地理位置和自然环境的优越性不言而喻。然而岛上地形起伏较大，山坡陡峻，大部分坡度达45° 以上，可建设基地狭长且复杂，这是一个极具挑战的项目，也是项秉仁实践关于建筑与环境共生的理念的绝佳机会。他希望设计能充分结合原有自然山水条件、地形地貌，力求使建筑自然生长于环境之中，实现建筑与环境和谐共生。

为了更好地顺应基地的环境特点，建筑总体布局、道路系统规划以及景观规划设计都做出了精心的考虑。由于度假酒店的功能庞杂，而整体基地又较为狭长，在确定酒店用地时，基地中部相对平缓，进深较大的地块自然作为酒店主体建筑区的首选落位，而西侧山地起伏区域则用作度假别墅组团的用地。酒店形体面向湖面舒展铺陈开来，并化解为多个体块功能，通过连廊连接，避免了在基地上形成庞大敦实的建筑体量。道路系统沿地形走势并结合建筑群体关系进行规划，为尽享岛上景观资源，沿湖低标高布置景观步道，连接规划中的多个景观节点，构建一个立体分层观景的休闲系统。尽可能保护原有绿化植被，保持岛屿原生态风貌，同时根据季节变化丰富配栽不同色彩的植物种类，形成物种多样的生态群落。结合地形穿插布置泳池、小品、亲水平台等人工景观，在满足休闲度假的同时，实现自然景观和人工景观有机渗透，充分融合。

在酒店的室内空间处理上，也尽可能将良好的自然景观引入室内。通高两层的大堂面向水面180° 观景，下沉式的早餐厅在享有室外美景的同时避免了对大堂景观视线的遮挡。酒店其他开放性公共空间如中餐厅、酒吧等均面景而设，使置身酒店的客人，闲庭之余处处游赏湖光山色。酒店客房以大堂为中心分设东西两翼，以最佳景观面排布多数客房，垂直湖景客房采用斜窗处理，以此扩展房间的景观视野。

酒店落成之后，建筑群体依山而筑，据地形变化而高低错落，表达了对自然地貌的尊重。建筑形体逐层退台跌落，消解了建筑体量。立面墙体色彩贴近自然，结合坡屋顶形式，更好地融入自然山水之中。当我们揽湖远眺时，在青葱山林植被的映衬下，建筑轮廓与岛屿天际线和谐和声，体现了项秉仁在设计之初所倡导的建筑“自然生长”的理念。

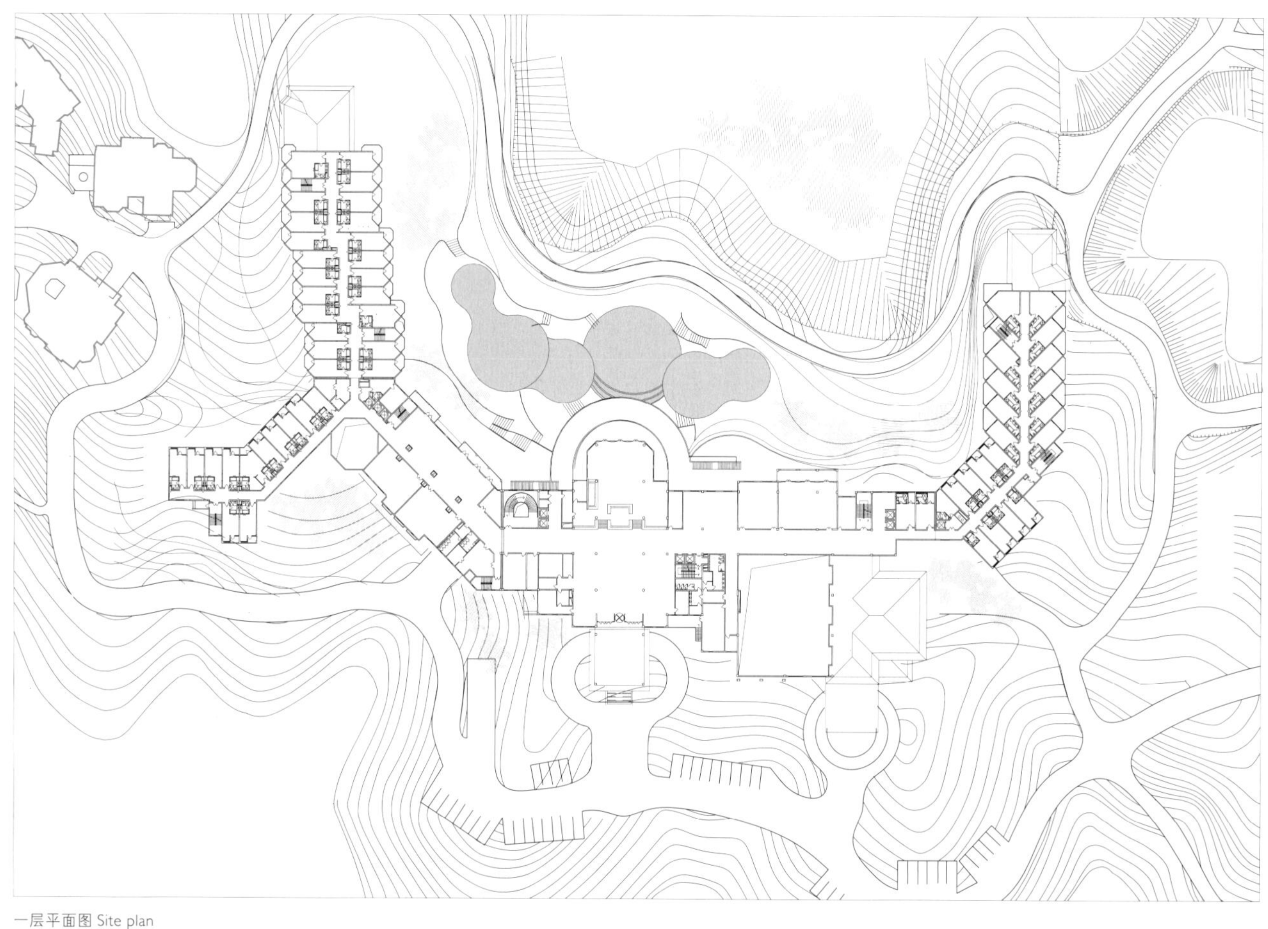

一层平面图 Site plan

N

Qiandaohu Runhe Jianguo Hotel, located on an east-west peninsula called Menggu Island in Qiandaohu, is the project with the most complicated topography that Xiang Bingren had ever encountered. The long and narrow Menggu Peninsula extends all the way to the central area of Qiandaohu Lake. The island's south, west and north are surrounded by water, with the east end directly linked to the urban roads leading straight to Qiandaohu town. The island's beautiful scenery, lush vegetation, geographical location and outstanding natural environment are renowned. However, the irregular shape of the island and steep slopes (over 45°) provided a seemingly impossibly narrow and extremely complicated usable area for the hotel. This was a very challenging project and an excellent opportunity for Xiang Bingren to explore the concept of symbiosis between architecture and the environment. He hopes that the design can fully integrate the original landscapes and topography as if buildings naturally "grow" out of their unique environments, achieving a harmonious coexistence of architecture and nature.

To better respond to the environmental characteristics of the site, Xiang Bingren carefully considered the overall layout of the building, the planning of the road system and the planning and design of the landscape. Due to the complicated programmatic composition of the resort hotel, and the relatively narrow shape of the site, when Bingren was determining the layout, he chose to locate the main building of the hotel on the central piece of the island where the land is relatively flat while dedicating the hilly west mountain lands for the construction of villa clusters. The main part of the hotel faces the lake and stretches in form to obtain a relatively long landscape surface. It is divided into several blocks with different functions connected by corridors to avoid an excessive architectural footprint. The road system respects the topography and plans according to the relationship between the building groups.

# Qiandaohu Runhe Jianguo Hotel

尊重场地，匠心思考 Night view of the resort

To make full use of the island's landscape resources, the design team has arranged a near-water landscape trail along the lake and connected multiple landscape nodes to construct a three-dimensional viewing and leisure system. In order to protect the original green vegetation and protect the natural ecology of the island, the design team configured the plants with different colors according to the seasonal changes so as to form a variety of ecological communities. In combination with the change of the terrain, the design scheme scattered landscape elements such as a swimming pool, landscape pieces and a water platform across the island to achieve an interweaving of natural and artificial landscapes while fully satisfy the demand for leisure spaces.

The interior design of the hotel also aims to provide as much natural landscaping to the inside as possible. The two-story lobby includes a 180° view of the water, while the sunken breakfast room offers outdoor views while avoiding obstructing the lobby's view. The hotel's other open spaces, such as Chinese restaurants, bars and so on are all set in front of the scene, so that guests staying in the hotel can enjoy the surrounding lake and mountains unobstructed. The hotel rooms are located on the east and west wings of the lobby. Most of the rooms allow sites of beautiful scenery with a sloped window to allow for an optimized viewing angle.

According to the topographic changes in the hotel complex, the buildings are positioned according to the natural landscape. Terraced designs, neutral color and sloping roofs of the building all help to break down the volume, and to better integrate the building with the surrounding landscape. When we look over the lake from the building, against the backdrop of the green vegetation, the architectural outline is in harmony with the island skyline. This reflects the concept of "natural growth" of the architecture advocated by Xiang Bingren at the beginning of the design.

# 评论
Review

那一年，酒店开幕，项老师与我受邀前往。记得那天晚霞很美，千岛湖山色水光，别样的诗情画意。次日清晨，我们散步到东侧的山顶公园。上山的台阶齐整，透着细碎阳光的林木生意盎然，一路上几乎没什么人。途中不时望向西侧，在树缝中偶然会瞥到酒店的一角。走走停停，在木华芬芳中蜿蜒上到山顶的庙里。庙里供奉文曲星，有几个长辈带着小辈前来礼拜。上到阁楼，往外眺望，就看到酒店的全貌。清晨的微风略有凉意，项老师和我倚着栏杆，静静地看着湖水松林环绕的酒店在晨光中伸展着，和着光线的变化，似乎有生命地呼吸着。

八年的时光，从人迹罕至的荒坡半岛，如今成为一处度假的栖居场所，承载人们的欢欣喜悦，其中的不容易迟早会化为美好的回忆，淡淡地无从提起。但此刻俯瞰拥抱着湖光山色，沐浴于水汽氤氲中，必然历经寒暑风霜的这一酒店，于项老师心中油然而生的或许是一种创造者的爱怜之情。而我，深感幸运。

蔡沪军

上海彼印建筑设计咨询有限公司创始人、总经理、总建筑师

I still remember the beautiful sunset, shining above the Qiandaohu Lake quite poetically, on the day Mr. Xiang and I attended the hotel's grand opening. The next morning, we strolled to the Peak Park on the east side but are surprised that we barely saw anyone else along the way. The steps up the mountain were well maintained, and the woods bathed in sunlight were full of life. From time to time, as I looked to the west side, I could see corners of the hotel in the crack of the tree-crown. After a few stop-and-gos in the fragrance of mountain flowers, we made it to the temple on the summit. The temple is dedicated to Wenquxing, who was visited by several elder worshipers. Going up to the attic and gazing out, my eyes were full of a panoramic view of the hotel. Mr. Xiang and I leaned against the railing and quietly watched the tree-enclosed hotel in the slightly chilly morning breeze, as if the hotel was breathing and changing with the daylight.

Eight years later, the once inaccessible desolate peninsula is now a vacationland of joy. The transformation was not easy, however, the painstaking work it took to realize this ambition seems impolite to mention today. But at this moment, overlooking the lake and the mountains, bathing in the mountain mist, Prof. Xiang, I assume, must have felt a tender affection as creator of the project, which is bound to go through rain and wind. For me, in that same moment and space, I simply felt lucky to be there.

Cai Hujun

Founder/ General Architect/ Chief Architect, Being Studio Architects

2010 / 2012

# 招商·万科佘山珑原别墅

佘山珑原别墅是位于上海佘山风景区内的一处高档别墅区的规划设计项目，由招商、万科两个品牌共同开发，产品涵盖双拼别墅、类独栋、类双拼联排别墅和叠加别墅4种类别。项目基地位于松江区广富林遗址公园北侧，毗邻国家4A级景区资源，周边用地均为别墅及文保用地，享有得天独厚的先天地理优势。

在设计之初，适逢新中式刚刚兴起之时，别墅类建筑风格开始从美国梦和欧陆风慢慢回归本土，传统文化与商业地产之间正寻求结合点，雅致、含蓄的中式情节逐渐被提炼起来。加之上位规划对于此项目的风貌要求，项秉仁认为这将是一个有别于常规商业地产开发的项目，中式传统人居理念将得以体现，他希望设计能以传统中式居住形式为借鉴，满足当代居住需求，演绎现代中式居所，打造高品质的现代生活。

项目整体空间规划传承传统中式居住聚落空间，形成特点鲜明的“街”“巷”“院”三级外部空间结构。其中，“街”道空间对应社区公共活动的开放空间；“巷”道空间对应居住组团中邻里交往的公共空间；“院”落空间对应居住空间中的私密空间。以“院”为中心的居住体验模式成就了其阔院别墅的特色。前院是生活空间的核心，侧院及中庭是室内生活空间的延伸，空中庭院是绿色空间的立体延伸，公共的庭院则是社区交流的最佳场所。

同时在空间结构上借鉴了传统江南建筑的空间形态特征，在序列上强调“围合”和“轴”–“进”的变化；在形态构图上强调“对称”中的“不对称”以及“间”作为基本的构图原则；在形体要素上强调对传统建筑中“墙”“门”“窗”“檐”等建筑构件细部的现代转译表现。突出“墙”在公共空间和私密空间中的限定作用，形成内向型的居住空间，营造传统中式居所特有的场所精神。总体景观规划遵循空间规划的脉络，尊重场所特征，并吸收传统园林“步移景异”“小中见大”“师法自然”等设计手法，以现代造景技法表达传统中国意蕴。

建筑立面素雅而精致，主色调为白色及深灰色，体现传统水墨江南的意蕴，注重细节的同时抹去了不必要的装饰，石材及金属、玻璃替代了传统的白墙、木作及窗户，体现出现代建筑的特性。建筑并没有刻意地模仿传统建筑，而是在收放之间，在简练的手笔中有了传统的神韵，在传统意境中呈现出不失当代气息的建筑形象。

相地原生，师法自然的人居环境 View of the clubhouse

Sheshan Longyuan Villa, within the Shanghai Sheshan scenic area, was developed by a joint venture of CMBC and Vanke, and included housing typologies for duplexes, single-family houses, townhouses and overlay villas. The site of the project is at the north of Guangfulin Historic Park in Songjiang District. It is adjacent to a national 4A grade scenic spot, with surrounding lands allocated for villas and cultural preservation, which provide the project with unique natural and geographical advantages.

At the beginning of the design, the Neo-Chinese style had only just emerged along with a change of market preference from Western back to a Chinese style. Developers were seeking a way of combining traditional culture and commercial real estate. Hence, the elegant and implicit quality is highlighted. In addition to the planning requirements for this project, Xiang Bingren believed that this would be a project very different from conventional commercial real estate development and would reflect the traditional Chinese concept of human settlement. He hoped that the design could use a traditional Chinese style of residence as a reference while meeting the needs of contemporary living, creating high-quality modern life.

The spatial planning of the project inherits the characteristics of traditional Chinese settlements, forming a distinct structure of outdoor spaces that fall into three categories: "street", "lane" and "courtyard". Among them, the "street" space corresponds to the open space needed

风华清雅的建筑单体 Townhouse exterior

for community public activities; the "lane" corresponds to the space of the neighbors' communication; and the "courtyard" space corresponds to the private space of the dwellings. The "civil courtyard" at the center of the living model has become a central characteristic for the villa, with the front yard being the center of living space and side yards to accommodate outdoor activities. Also, the design of the vertical courtyard was meant to expand greenery into a three-dimensional experience. Moreover, the shared courtyard makes a great communal space.

At the same time, the design of the spatial structure also borrows heavily from traditional Jiangnan dwellings, emphasizing the concept of "enclosing", "axis" and "layer". The idea of asymmetrical elements within an overall symmetrical system and the concept of "panel" have been collectively deployed as the planimetric composition principle of the design. Moreover, characteristic formal elements from traditional buildings such as "wall", "window sash", and "eave" have been reinterpreted into a much more modern expression even as their aesthetic and spatial essence were kept intact. The "wall" is typically used as an instrument of dividing what's private and what's shared. Here it forms a courtyard-centered introverted living space, which echoes Chinese living philosophies and traditions. The landscape design follows the spatial layout, respects the different characteristics of different places and learns from the representative design methods of traditional Chinese gardens to create pathways with constant-

街巷院的设计理念 Design strategy of "street, lane and courtyard"

ly changing scenes. The effect is to imply that the universe within man-made boundaries must also learn from the nature. It succeeded in creating a modern landscape design that represents the traditional Chinese cultural connotations.

The building facades are elegant and delicate, with main colors of white and dark gray, reflecting the essence of the traditional Jiangnan dwellings. The design, while paying attention to detail, works to erase unnecessary decorations while replacing the traditional white walls with stone curtain-walls, traditional wooden structure with metal, and wooden sash with glass curtain-wall. The lasting effect is to reflect the characteristics of modern architecture. The design does not deliberately imitate traditional architectural styles, but still possess a traditional charm, showing the contemporary architectural image in the traditional artistic conception.

## 评论
## Review

珑原别墅是结合广富林的历史环境打造的一处现代中式别墅群，它将中国的人文特质融入到当代建筑中，在传统和现代生活之间架起了一座桥梁。
在具体的设计语境中，较于当时盛行的符号化中式，它是一次特别的尝试，项先生提炼出传统的语义，用最简练的建筑构成表达出了中式特性，寥寥数笔之间，建筑传达出明确的现代性，用抽象的方式完成了中式的使命。
抽象的美是能经过时间的考验的，多年后的今天再回看，它仍然能让人感到直抵心灵的力量。

肖俊瑰

上海天华设计总监，武汉天华副总建筑师

意境为神，中国文化传承的体现；本体为形，现代居住理想的承载。
在一山一址（佘山、广富林遗址）之下，项先生将文化回归与传承开创并置，将根植于江南的素雅与宁静融入会呼吸的意墅空间，得到的不仅仅是街·巷·院、黑·白·灰、水·桥·房，佘山珑原项目的设计带给我们更多的是对现代中式生活及意境的联想：

寻珑原
一路经行处，翠竹映熙纹。
白云依静渚，夏花闭闲门。
过雨望佘山，随源至先痕。
溪花与禅意，相对亦忘言。

郑滢

天华集团品牌公关总监、建筑委员会秘书长

The Longyuan Villa is a modern Chinese-style villa complex integrated into the historical environment of Guangfulin and combines Chinese perspectives with contemporary architecture thus bridging the gap between traditional and modern life. It is a unique attempt toward this goal and far different from the symbolic Chinese style prevailing at that time. Mr. Xiang extracted traditional Chinese semantics and expressed them with the most succinct architectural composition, imbuing the buildings with a strong yet abstract modern Chinese look. The beauty of abstraction here appears timeless. Looking back today, many years after the project's completion, the design can still shake you to your core.

Xiao Jungui

Design Director, Shanghai Tianhua; Vice Chief Architect, Wuhan Tianhua

The artistic conception constitutes the spirit of the project, expressing its Chinese cultural inheritance; the form of the project represents an ideal way of modern life. At the foot of Sheshan Mountain, upon the site of Guangfulin relic, Mr. Xiang juxtaposes culture and inheritance, integrating the elegance and tranquility rooted in the Jiangnan region into the living and breathing space of Longyuan Villa, creating not only lanes, courtyards, black, white, gray, water, bridges, houses, but also the artistic imagination of a modern Chinese lifestyle.

*Searching for Longyuan*
*Along the sunshine trail, I see bamboos casting mottled shadow.*
*A cloud low on quiet islet, summer blossoms sweeten a door in idle.*
*In Sheshan Mountain after rain, a brook led me to a site of remains.*
*Mingling with truth among flowers, I have forgotten what to say.*

Zheng Ying

Brand PR Director/ Secretary General of Architecture Committee, Tianhua Group

主入口透视图 Main entrance perspective

# The Six Villas, Wangshan



2011/

# 旺山六境

旺山六境是一次未建成的实践，它缘起于2010年的秋季，是一次带有乌托邦色彩的集群设计实践。业主希望在苏州市著名景区旺山度假村内打造六栋具有创意的低密度居住建筑，于是在苏州大学刘晓平教授的牵头下，六位国内外知名建筑师参与到这次实践中。项目基地坐落于半山，背靠一片精致的竹林，向南可以眺望山脚下的风景，颇有些气象。业主对设计师的要求十分宽松，只要满足面积需求和用地范围，可任由个人发挥。项秉仁希望在这次特殊的设计实践中，能融入他对于东西方居住模式和生活理念的思考。

旺山是苏州传统意义上的“郊野”，具有得天独厚的自然环境和悠久的士人隐居传统。而在中国改革开放以来的快速城市化中，旺山又被稠密的都市区所包裹，成为“城市”的有机组成部分。如何在这样一个兼具“市”与“野”特征的场所中塑造一种理想的栖居？项秉仁希望能够从中国文化中的栖居哲学中找到启发。所谓“大隐隐于市、小隐隐于野”的“市”与“野”是中国人对于居住的两种截然不同的意境所在，它们反映了两种栖居的态度。居于“市”即住在城市中，在一种嘈杂的环境中，需要以封闭的物理环境（院墙）去隔绝外部，而内部的建筑排列紧密。生活发生在院、廊、厅、堂、庭的小天地中。它与外部绝缘，却在内部创造出园林空间、礼序关系，是内向性的。然而居于“野”则全然不同。“采菊东篱下，悠然见南山”的生活方式所对应的栖居格局是自由、零散、开放的建筑布局，以求充分地与自然环境融合互动，它是外向性的。在具有“市”与“野”的双重特征的基地上，如果能有机会将这两种情境下的栖居方式融汇到一座居住建筑中去，并结合现代生活品质的要求，这将会是一次非常有趣的尝试。

有了这样的思考，项秉仁开始解构当下理想的居住方式的组成部分。我们知道，住宅建筑空间素来就有动静的双重属性。一方面，客厅等公共区域需要能提供更好的聚会、交流等活动场所，而卧室等区域则需要私密性和更幽静的景观。设计将住宅中各种较为公共、开放的空间组织在较为内敛的平面中，称之为“市”的居所，它由门厅、客厅、餐厅、会客厅、茶室等公共空间围绕着一系列微型苏州园林展开，所有这些空间都为外部的围墙所围合，几乎没有外窗，形成充分的内向性的建筑首层。别墅的四个卧室空间安排在二层，成为视野极佳、毫无拘束的玻璃盒子，它们有机地与屋面绿化和周遭的环境互动、融合，并且通过一条空中步道联系起来。步道蜿蜒曲折并延伸到建筑背后的竹林中，使主人足不出户便能完成一次空山新雨后的漫步。所有卧室都可以得到360° 的观景视角，实际上它们更像是散布在山林中的几顶帐篷。这是对外向性的

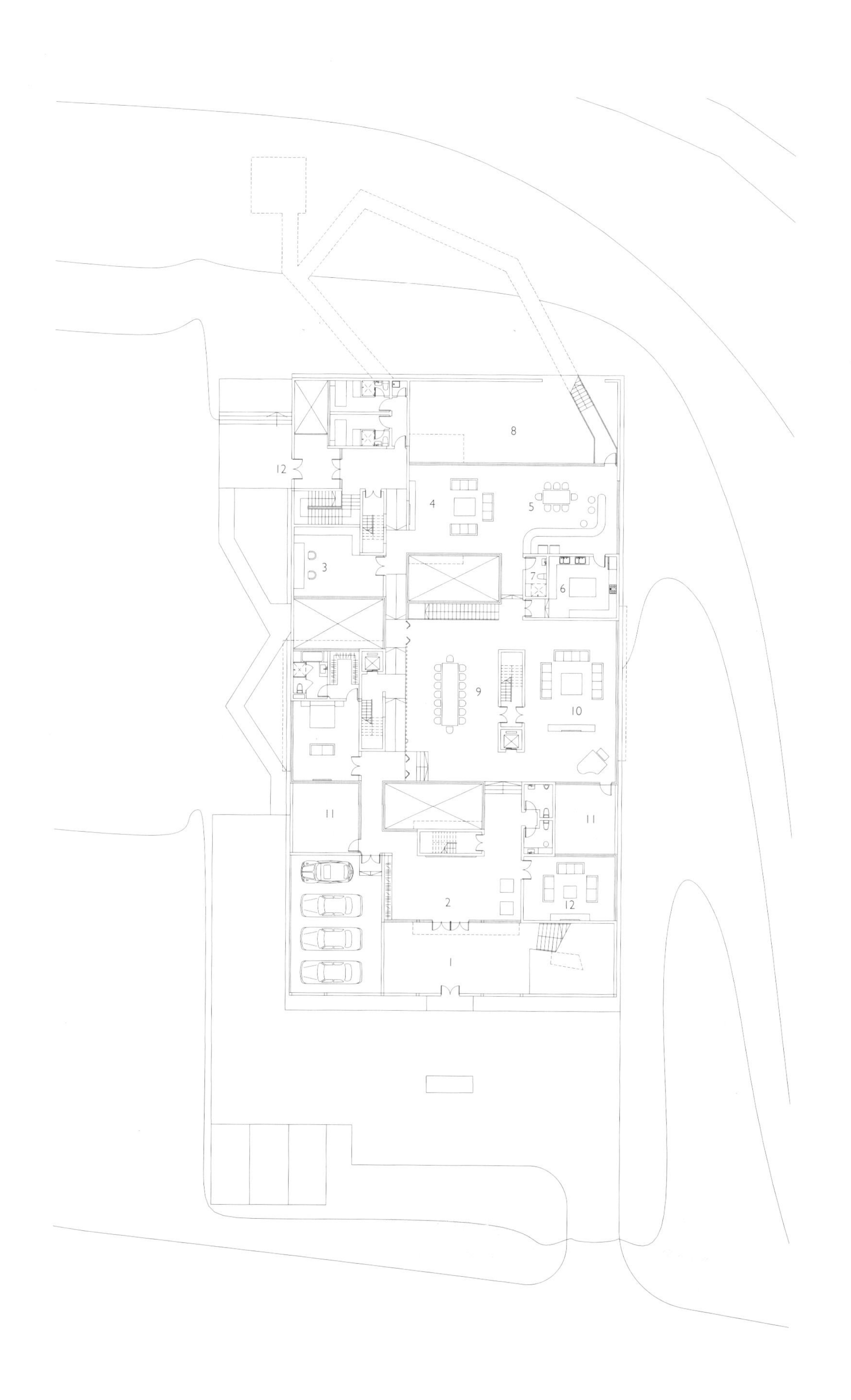

一层平面图
First floor plan

1 院子
2 门厅
3 书房
4 家庭室
5 吧台
6 厨房
7 卫生间
8 莲园
9 餐厅
10 客厅
11 花园
12 会客厅

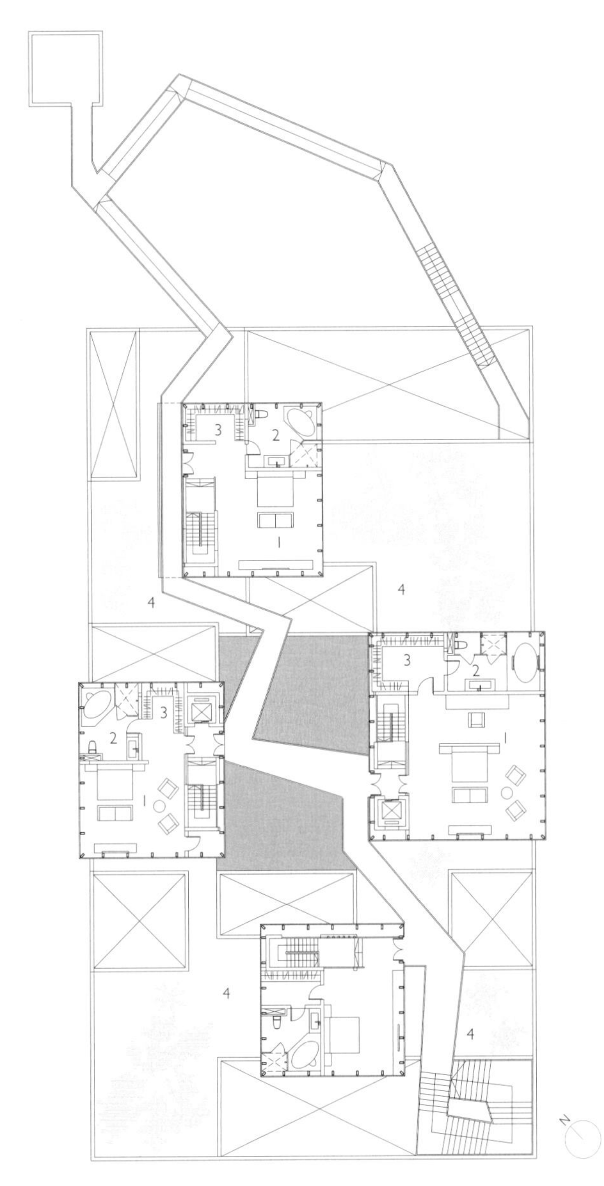


二层平面图
Second floor plan

1 卧室
2 卫生间
3 衣帽架
4 梯田

"野"的居所的理解。

虽然这个作品由于种种原因并未最终建成，但我们仍能为整个思辨的过程而感到雀跃。这是一次关于居住建筑类型学创新的试验。"市"与"野"的居住智慧在设计中形成了和声，而相应的建筑空间也因此变得富有戏剧性。同时，我们也隐约感受到，项秉仁借着这个机会，用他的设计表达出对于现代主义建筑前辈的无限敬意，并试图证明在当今的经济和科技条件下，以往被指责的现代主义建筑的一些弊端是完全可以避免的。时至今日，过去遭人诟病的密斯式的玻璃盒子住宅完全可以通过多层中空玻璃、雾化技术和智能控制被优化，而当初现代主义建筑大师的理想在今天依然可以绽放其生命力。

The Six Villas, an uncompleted project begun in the fall of 2010, was a cluster design with Utopian characteristics. The client hoped to build six innovative, low-density residential buildings in the Wangshan Scenic Area of Suzhou City. Under the lead of Professor Liu Xiaoping of Suzhou University, six renowned architects participated in this project. This project was located on a hillside backed by exquisite bamboo forests and provided a great location on the south to overlook the scenery at the foot of the mountain. The client had loose requirements for the architects, giving them rein to imagine and integrate their personal preferences as long as all site restrictions were met. Xiang Bingren hoped to integrate his understandings of Eastern and Western residential patterns

平台景观 View of the terrace

and lifestyles into this special design project.

Wangshan, as a suburb of Suzhou City, contains unique natural environments and a long tradition of seclusion. However, under rapid urbanization following the economic reform, Wangshan is now surrounded by dense metropolitan areas. Wondering how to build an ideal residence in such a place that combines the characteristics of both the "city" and the "wild", Xiang hoped to find inspiration in Chinese culture and philosophy. The old saying "Small-time hermits hide in the mountains, yet the greatest hermit will retreat into the noisiest fair" depicted two completely different concepts and attitudes of living within Chinese culture. Living in the city requires a compartment, such as a courtyard wall, to separate the residential area from the noisy urban environment, while the interior structures are arranged closely. Life in the little world within the courtyard, porch, hall, and atrium is isolated from the outside, creating a private garden space, courtesy and etiquette inside. On the contrary, living in the wild is completely different. The lifestyle depicted in the classical Chinese poem "while picking chrysanthemums beneath the eastern fence, my gaze falls leisurely upon the southern mountain" is free and leisurely, corresponding to the open, scattered architectural layout in order to facilitate prolonged and constant interactions with the natural environment outside. At a place that bears characteristics of both the "city" and the "wild", it would be a very interesting attempt to

incorporate the demands of high-quality modern life and combine these two different lifestyles into the design of one residential house.

With such thoughts in mind, Xiang began to deconstruct the components of ideal modern living patterns. As known, residential spaces always have dual attributes, static and dynamic. On the one hand, public spaces such as the dining room should provide a space suitable for gathering and communication, while personal space such as bedroom rooms should be more private and quieter. Xiang's design included a lobby, living room, dining room, lounge, tearoom and other public space surrounding a series of miniature classical Suzhou gardens. All these spaces on the first floor were separated from exterior environment with courtyard walls with few windows. The four bedrooms are on the second floor, formally, work as four glass boxes interconnected by a walkway, each with extraordinary views and interacting with the roof greenery and surrounding natural environment freely. The walkway is twisted and reaches into the bamboo forest behind the building, enabling the owner to enjoy a walk after a fresh rain without leaving home. All bedrooms offer a 360-degree sweeping view, like tents scattered in the forest, reflecting Xiang's understanding of the "wild".

Although the project was never realized for a variety of reasons, the entire planning process was still full of fun and excitement. This was an innovative experiment in residential typology. The concepts of the

^ 餐厅透视图 Dining room perspective
v 客厅透视图 Living room perspective

"city" and the "wild" coincide harmoniously in this design, accordingly, resulting in rather dramatic spaces. In the meantime, Xiang took this opportunity to pay homage to the modernist, and tried to prove that formerly criticized drawbacks of modern architecture can be completely resolved with the support of current economic and technological developments. At present, the previously criticized Miesian glass design can be optimized using multi-layer insulating glass, atomizing technologies and thoughtful control. The ideals of modern architecture masters in the past can still shine and radiate today.

## 评论
## Review

作为旺山六境国际集合设计的策划者，我从项教授的构思过程中，看到他经历了从物（形态和空间），从外在（基地）出发，纠结反思后最后回到"情境"和"心境"，回到"心"和"意趣"的过程，也就回到了文人造园的作者本位，这是项教授对我在项目中设定的"传统与当代，全球化与本土化"的一种回答，项教授这种源于现代主义精神的主体性创作思维，在我看来是远高于贴着各种时髦标签，但却"无我"的设计明星。

方案实现了对市与野两种情境的巧妙复合：首层"造园"封闭的院墙围合之中，密布院、廊、厅、堂。营造了层次丰富、引人入胜、小而不尽的"园林"生活空间。二层的自由零散的玻璃盒子（卧室）则最大程度拥抱山林环境的盎然野趣。这个方案实际也是东西方两个经典住宅样本的结合，首层的园是中国人居精华，二层的西方现代经典玻璃盒住宅也按游园方式布置，两者融合，毫无违和感。这体现了项教授学养深厚，融会贯通，游刃有余。

另外，不同于玩票建筑师的试验品，方案的空间布置与生活内容完全是舒适齐全，品质高贵的，这体现了项教授作为杰出职业建筑师的深厚功底。作为建议，在我看来，下层公共区、上层卧室区这个分区设定也可以适度打破混合，那么生活的故事将会有更加丰富的情境。

刘晓平

苏州大学建筑学院硕士生导师
华东都市建筑设计研究院副总建筑师

As the master planner of the Six Villas in Wangshan, I witnessed Prof. Xiang's entanglement of starting from physical considerations (form and space), then moving on to the context (site), before finally returning to concerns for "joy" and "mind". This process of "coming back" coincides with the process of returning to the author's standard of literati's garden making, which is Prof. Xiang's response to the theme I set for the project: "traditional yet contemporary, global yet local". Creative thinking that stems from the spirit of modernism is far more valuable than those design celebrities who bear all kinds of fashionable labels but carry with them no soul.

The design realizes an ingenious combination of the two conditions of a healthy, well-balanced life: city life and rural life. By designating the first floor as a garden enclosed by walls with courtyards, corridors, halls and churches a small yet fascinating "garden" living space is created with rich layers. The second-floor works as a freely arranged glass box (bedroom) embracing the wilderness of the mountain to the greatest extent. This strategy perfectly combines two classic residential samples from the East and West. The garden on the first floor represents the essence of a Chinese way of living while the classic Western modern glass box houses on the second floor are arranged to mimic the experience of garden wandering. This reflects Prof. Xiang's knowledge and experience that allow him to integrate different cultural modes so effortlessly.

In addition, unlike experimental products of some part-time architects, the spatial layout and living necessities are complete, comfortable and of high-quality which reflects the profound knowledge as an outstanding professional architect. It does seem to me, however, that the division of the lower level as a public area and the upper level as bedroom could be blurred a little bit, to enrich life with more interesting scenarios.

Liu Xiaoping

Master Supervisor, School of Architecture, Suzhou University
Deputy Chief Architect, East China Urban Architectural Design & Research Institute

# 4

## 都市高层建筑：理性与冲动

高层建筑是城市建筑的一个重要类型，它对于城市天际线和城市形象有着至关重要的影响，一座有气质的现代高层建筑对于城市整体风貌的贡献是决定性的。自1998年项秉仁在南京玄武湖边设计了一栋现代典雅的椭圆形透明塔楼之后，他在全国各地主持设计了数栋高层建筑，这些项目从设计之初就在践行项秉仁的城市建筑理念，重视拟建建筑对于城市局部环境和整体形象的改变及影响。他遵从城市设计导则的指引和限制，以发展的眼光处理新老建筑的共生和延续关系，在尊重城市原有历史文化的前提下营造具有当地特征的当代建筑文化，以专业精神和对建筑细节的高度专注，创造一件件适应时代要求的建筑精品。这些设计思考维度上的丰富性，使我们体味到理性冷峻的几何外皮下的内在多元，感受到建筑物蓬勃的生命力。

## Urban High-rises: Reason and Impulse

As a main type of urban architecture, high-rise buildings constitute a good deal to a city's skyline and its image. It's not exaggerating when we say a temperamental modern high-rise building defines the overall style of a city. Xiang Bingren's first high-rise project was a modern oval transparent tower by Xuanwu Lake in Nanjing in 1998. Since then he has designed several more of this typology throughout the country. He takes the opportunity to exercise his ideas of urban architecture in these projects, paying much attention to the influence of the proposed building on local environment and overall image of the city. He sees the coexistence and continuation of new and old buildings from a sustainable point of view, and tries to create a contemporary architectural culture with local features on the premise of respecting the city's original historical culture. His professionalism and focus on details help create a number of excellent works that meet the requirements of their times. The richness of his design thinking reveals the inner diversity and vitality of buildings under their cold and rational geometric skins.

# Jiangsu Telecom Multi-functional Building

1998 / 2002

# 江苏电信业务综合楼

江苏电信业务综合楼这座椭圆形的建筑无论建筑空间还是建筑形态都成为当时令人耳目一新的开创性设计。20世纪末期，正值中国城市建设起步并逐步进入高速发展之际，设计、管理、施工等各方面因素使得这一时期的高层建筑设计尚处于摸索阶段，设计上并无太多成熟案例可循，建成项目品质也良莠不齐。项秉仁希望能在妥善处理好城市、建筑两者的基本问题的同时，提出具有时代精神的有新意的设计方案。

基地东临南京玄武湖和古城墙，西临中央路，南侧和北侧是已建成的高层和多层公寓楼。当时的南京城高层建筑并不多，隔湖相望，拟建的电信大楼将在相当长的岁月中孤身兀立在城市天际轮廓中，因此这座楼的单体形体设计，包括造型的单纯性、永恒性、各个方位的视觉形象和比例等问题都显得格外重要。我们能看到，塔楼有意与街道成45° 角偏转布置，这样在场地内划分出两个不同功能的广场——一个面对公共开放和一个面向内部使用，同时这样的偏转也为办公塔楼争取了最大限度的景观视野和朝向资源。简洁明快的椭圆形建筑体量设计，具有极强的区域标志性和引导性，照顾到城市各个方向的视角，圆润的外观属性与碧波荡漾的玄武湖相得益彰。

塔楼主体功能主要是办公和电信技术业务，由于企业自用的设定条件，项秉仁在内部空间设计中引入了当时国外较先进的空中生态花园的概念——塔楼每四层设计一个通高的生态花园，以调节内部办公环境和高空小气候。这种在塔楼内部引入绿色和公共交往空间的方式为当时高层建筑生态化设计的初步实践提供了宝贵的设计经验，也在实际运用中获得了良好的体验。建筑物外皮采用呼吸式玻璃幕墙，高透玻璃和银色镀膜玻璃配以金属装饰构件，突显出建筑的现代科技感。流畅高耸的椭圆主体，配合顶部造型以及空透的空中花园，产生出新颖别致的变化，使其在南京办公楼中格外引人注目。

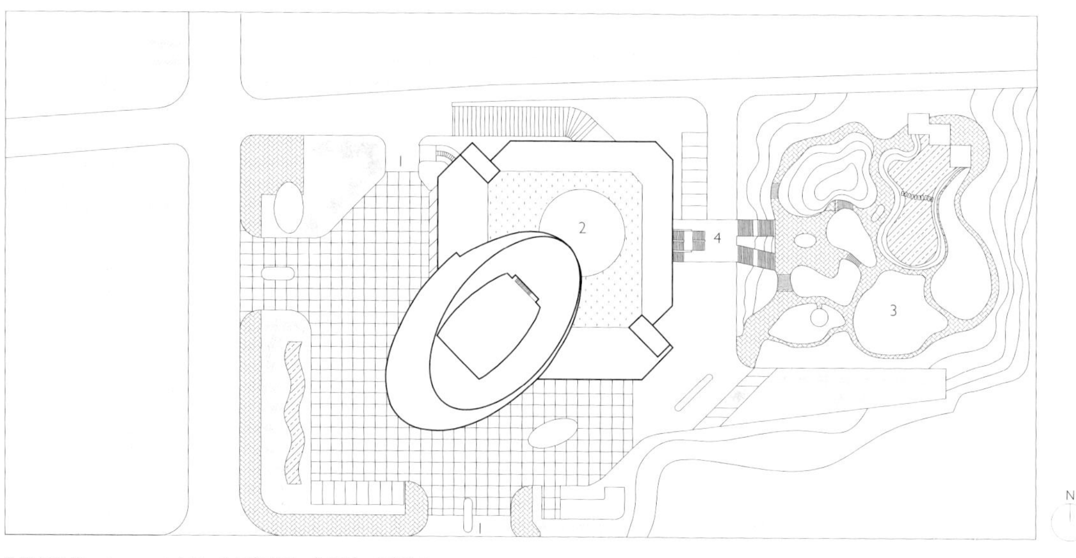


总平面图 Site plan　1 入口　2 屋顶花园　3 花园　4 建筑入口

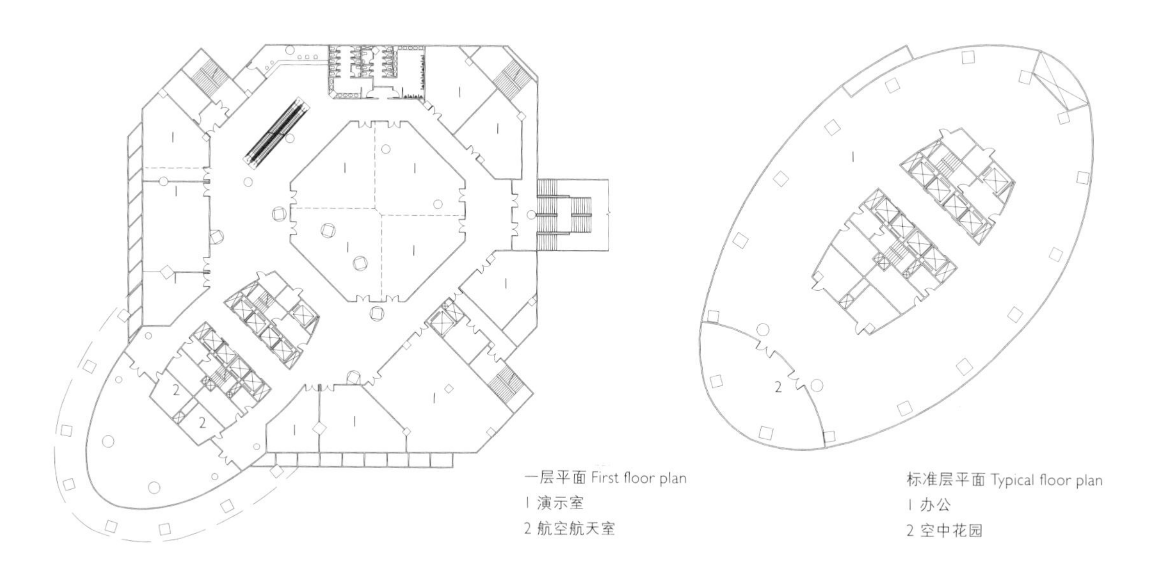


一层平面 First floor plan
1 演示室
2 航空航天室

标准层平面 Typical floor plan
1 办公
2 空中花园

Jiangsu Telecom Multi-functional Building, the peculiar design was quite refreshing for its innovation in both form and space. In the late 20th century, as China's urban development was about to accelerate, the country encountered a problem. Because of the immaturity of design, management and construction, high-rises building designs were still at an exploratory stage. Lacking many successful precedents to learn from, towers built at this period were of varying quality. Xiang Bingren was hoping to originate groundbreaking design ideas while properly dealing with the relationship between the building and urban context.

The site faces Xuanwu Lake and the ancient city wall on the east with Zhongyang Road to its west. North and south of the site are office towers and high-rise residential buildings. There were few high-rises in Nanjing city at that time, the new Telecom tower would be standing alone by the lake for quite a long time as a symbol of formal eternity and purity. Strong, thoughtful visual images perceived from all angles as well as proper proportion of the building were regarded as extremely important. We can see that the longer axis of the ellipse is at an angle of 45 degrees to the major street, dividing the site into two plazas: one facing the street with the other more internalized. This strategy also optimizes the view and natural lighting of the tower. The simple yet strong

# Jiangsu Telecom Multi-functional Building

模型照片 Model

oval-shaped floor plan is highly iconic and eye-catching with good visual homogeneity. The rounded form of the tower and the fluid scenery Xuanwu Lake compliment each other perfectly.

The tower was designed to accommodate mainly offices and telecommunication technological services. Due to the project's built-to-suit nature, Xiang Bingren introduced the idea of an ecological hanging garden, creating a double-height space every fourth floor to regulate the micro-climate of the tower and to improve the office environment. The introduction of greenery and communal spaces, on the one hand, provided valuable experience for the eco-design of high-rises which was something to be tested at that time, and on the other hand, amenities created by the design strategy are very much appreciated by users of the building. As for the building facade, a respiratory glass curtain wall was applied. The combination of highly permeable glass and its silver coating with metal decorative components presented a futuristic look. The Jiangsu Telecom building turned out to be a quite striking member of Nanjing's skyline, thanks to its soaring yet stream-lined form complimenting its special roof design and transparent hanging gardens and creating a non-conventional building rhythm.

实景照片 Night view

## 评论
## Review

位于南京中央路这一城市主轴上的江苏电信业务综合楼具有一种非常特殊的优雅气质。这种优雅的气质首先来自巧妙的建筑平面与形体。在玄武湖的清波之畔，圆形体量更适合这个场地，人工造物与自然之间的冲突与违和感得到缓解，同时从城市各个方向上都获得了较为均衡的标志性形象。与街道之间呈现的45°夹角是神来之笔，在兼顾最佳采光和景观朝向的同时，不着痕迹地完成了对于场地关系的清晰划分与界定：外与内、属于城市与属于风景。建筑优雅气质的另一个来源是精妙的材料和细部。立面上玻璃透明度的不同变化诚实反映出建筑内部的不同空间，反射率的精确控制进一步消解了建筑的体积感，又不至于过度闪亮炫目，建筑整体上完美融入到南京苍穹的晴雨晨昏中。

这栋建筑丝毫没有同类高层常犯的浮夸、躁动之弊，而是自信、平和、恰如其分。她以轻盈、明快的现代建筑材料和语言与这座城市的厚重历史、山水风华和谐共存，在建成近20年后也没有因时光流逝而失色。

华晓宁

南京大学建筑与城市规划学院院长助理、建筑系副主任、副教授、硕士生导师

The elegance of Jiangsu Telecom Multi-functional Building first and foremost comes from the skillfully conceived architectural plan and its form. A circular volume is definitely the most suitable choice for the site on Xuanwu lakeshore, resolving the conflict between artificial and natural entities, presenting an equally iconic image to all parts of the city. Moreover, the 45° angle between the building and the is ingenious as it both maximizes lighting and scenic view while subtly yet clearly organizing the site and defining the boundaries for both internal and external as well as urban-oriented and landscape-oriented spaces. Secondly, the elegance comes from the exquisite materials and design detail. Glass facades with transitioning degrees of transparency honestly reflect the various functions of the interior space. The reflectivity-controlled glass facade is subdued, causing the building to almost vanish into the rain and gloom of Nanjing city.

This final design resists appearing exaggerated or instigating as might be expected from a high-rise office building. Instead, it is self-confident, peaceful, and appropriate.

Hua Xiaoning

Dean Assistant/ Deputy Director/ Associate Professor/ Master Supervisor,
School of Architecture and Urban Planning, Nanjing University

建筑在城市环境中 Building in urban context

# Nanjing Telephone and Televison Service Center Building



1998 / 2002

# 南京电视电话综合楼

南京电视电话综合楼是与江苏电信业务综合楼同期设计建设的又一高层建筑。这栋建筑物位于南京市中心鼓楼市民广场北侧。基地一侧已建有一栋11层的电信业务楼，业主希望新建一幢高30层、总面积约40 000平方米的塔楼以满足其业务扩张的需求。在项秉仁看来，已有的建筑不应进行过多的改造，而应该作为南京城市发展的印迹以表达历史演进；新的建筑则应设法与其产生对话，在视觉上形成呼应，但又不能只是简单地复制立面。

如何构建新建塔楼和已有业务楼之间的对话成为设计伊始首要解决的问题。项秉仁通过四层连廊的设计，来协调新旧不同尺度的建筑，也在功能和形态上起到了相互联系的作用。从城市设计的角度看，这样的处理也有助于新建建筑对市民广场空间的围合和视觉上的应答。同时，借助动态均衡的整体构图和实体、色彩、层高、尺度的一致来达到新老建筑视觉上的和谐共生。新建塔楼以向上挺拔的姿态，和老楼敦实的体量形成对比，体块的切割变化隐喻着尺度的某种内在关系，以此产生对话。

除了处理新老建筑之间的对话关系以外，项秉仁对新建建筑的肌理也进行了推敲。很显然，新建的高层建筑本质上较之前并无太大不同，若要着力体现时代特征就需要运用当下新型的建筑材料。于是，塔楼主体应用多种新型幕墙和装饰制品，除石材幕墙以外，还涉及多种玻璃，如彩釉玻璃、镀膜反射玻璃、透明玻璃，等等。通过表皮的变化，塔楼外形丰富多变，构造出挺拔、轻盈、透明并富有纹理、层次的现代外部形象。在靠近市民广场的裙楼部分，通过通透的连廊以及活泼的外墙金属框架增添亲和力和趣味性。细腻的建筑立面使整个建筑在硬朗中透出活跃的气息，充满了时代感和现代感。

模型 Model

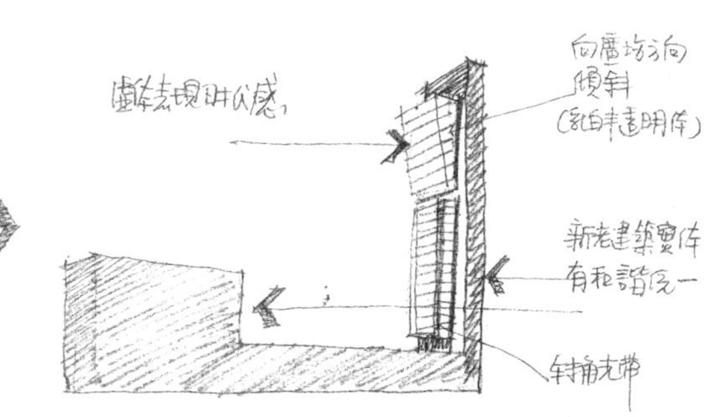


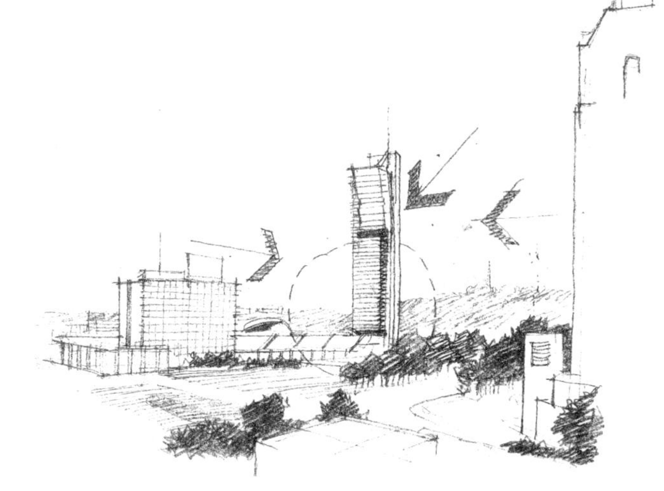

手稿 Sketches

The project for the Nanjing Telephone and Television Service Center is a contemporaneous high-rise project of the Jiangsu Telecom Multi-functional Building. The site is located on the north end of Gulou Civic Square in the very heart of Nanjing City with an existing 11-floor telecom building. The client wanted to build a 30-floor tower with 40,000 square meters of total floor area to fulfill the needs of their expanding business. Xiang Bingren believed that an overall reconstruction of the existing building wouldn't be necessary and that something should be kept intact as part of the narrative of Nanjing City's historical revolution. It was understood that while the new construction should produce a dialogue with the old, it was most important to create a visual coherency that avoids becoming a mere copy of the facade.

The key question at the beginning of this project was how to create an active dialogue between the new tower and the old one. Firstly, Xiang introduced a 4-floor interconnecting corridor to balance the striking difference of scales between the old and new and to create a visual and programmatic linkage between the two. It was also a very effective urban design strategy in that the corridor helped to define the edge of the civic square and make the new tower more visually responsive towards the square. Moreover, the harmonious coexistence of the old and

# Nanjing Telephone and Televison Service Center Building



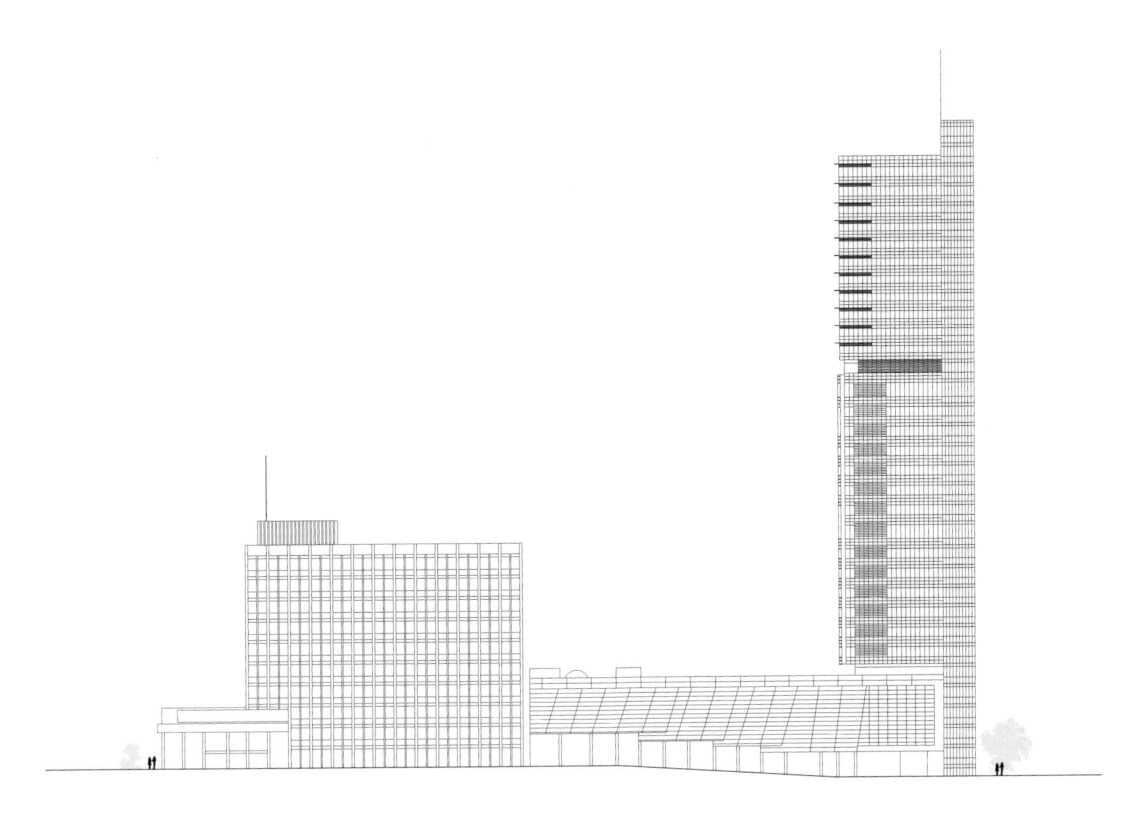

立面图 Elevation

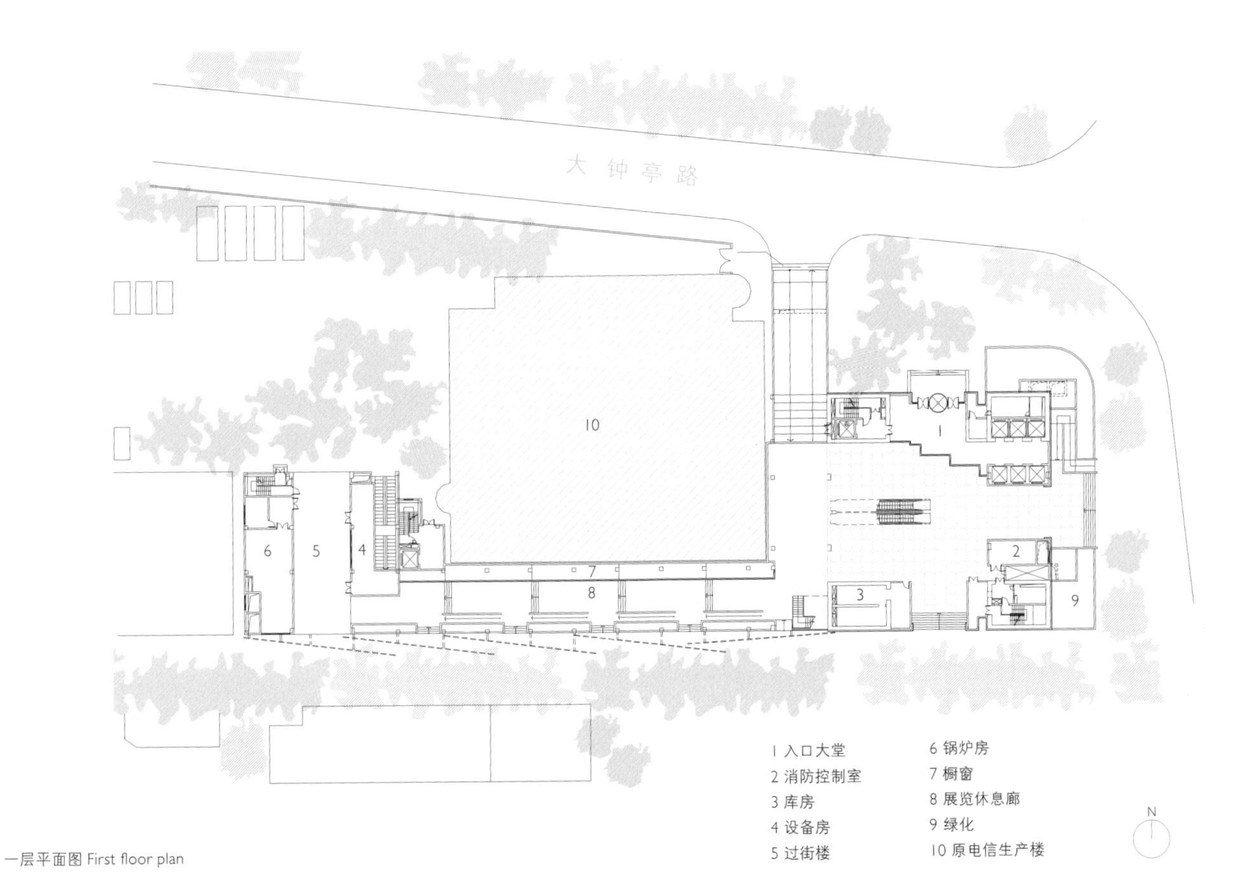


一层平面图 First floor plan

建筑局部 Building details

new is achieved through unified volume, color, story height and scale, as well as dynamically balancing the overall composition. The soaring figure of the new tower is in stark contrast to the stocky volume of the old, while the incision of the tower volume implies an internal relationship. In this way, a dialogue is created.

Apart from the concern of the old and new, Xiang Bingren also paid close attention to the texture of the tower's facade. Apparently, programmatically speaking, office towers are often quite generic with a strong tendency to avoid innovations. Aesthetically speaking, to convey the spirit of our time, the application of new materials is quite important. Based on this intention, various cutting-edge curtain walls and ornamental products were chosen. In addition to the traditional stone curtain wall, a wide range of glasses were introduced: enameled glass, coated highly reflective glass, transparent glass, etc. Variations on the facade led to a fascinating appearance of the tower that is soaring, light, transparent and ample with details and visual layers. The special design of see-through corridor with metal frame for the podium facing the civic square has tremendously increased the visual and spatial interest and affinity of the building. The exquisite facade has successfully endowed the towering and austere figure of the tower with a sense of livelihood and modernity.

# 评论
# Reviews

南京电视电话综合楼是南京鼓楼电信大楼的改扩建工程。11层的鼓楼电信大楼是中国七八十年代预制装配式建筑的重要代表，在南京市民的心目中也有着特殊的地位。无疑，与原电信大楼的关系是新建筑设计需要解决的核心问题之一。

设计采取的策略是不对原电信大楼的主体造型做任何改变，让它成为城市重要的记忆传承。新建的30层塔楼刻意布置在场地最东端，与原电信大楼拉开距离，避免庞大体量对原有建筑的不利影响。新建筑采用挺拔向上的柱状形体，与原有建筑的板式体量形成鲜明的对比，清晰肯定地表达出时代发展的印迹。另一方面，通过对形体的切削和不同材料的使用，将塔楼处理成若干体块的组合，化整为零消解塔楼的体量，在尺度层面与原有建筑和城市空间产生对话和呼应。塔楼面向鼓楼市民广场的部分，立面主要采用玻璃和金属材质，有效柔化了与城市公共开放空间的关系，化解了对广场的压迫感。塔楼东侧的实体部分则运用了与原电信大楼相似色调的黄色石材，与原有建筑形成了饶有兴味的“纵一横”呼应关系。

在这个项目中，项秉仁老师显示出了娴熟的技巧和审慎的态度，以“和而不同”的方式处理新旧关系，既不重复拷贝，又不标新立异。新建筑不仅与原电信大楼，更与鼓楼广场周边其他建筑以及广场空间本身和谐相处，创造出令人印象深刻的城市场所。

华晓宁

南京大学建筑与城市规划学院院长助理、建筑系副主任、副教授、硕士生导师

The Nanjing Telephone and Television Service Center Building is a result of the renovation and expansion project of the Nanjing Gulou Telecom Building. The 11-story Gulou Telecom Building serves as an important example of prefabricated buildings in China from the 1970s and 1980s. It also held a special position in the memories of Nanjing citizens. The relationship with the original telecom building was one of the core issues that needed to be addressed in the design of the new buildings.

The design strategy was not to alter the exterior of the original telecom building but to maintain it as an important heritage site for the city. The newly built 30-story tower was deliberately placed at the easternmost point of the site, separated from the original structure to avoid overshadowing the much smaller building. The new tower is tall and straight, contrasting greatly with the slab volume of the old building, expressing forcefully the passage of time. Moreover, through the subdivision of building volume and the use of different materials, the overall mass of the tower is broken down in order to create dialogue and harmonize with the existing building and urban context in terms of scale. The facade facing the Gulou Civic Square is composed primarily of glass and metal, effectively softening its relationship with the open, urban space beyond and resolving its oppressive dominance over the square. The facade on the east side uses yellow stone similar to the original telecom building, which forms an interesting “vertical-horizontal” effect, echoing between old and new.

In this project, Mr. Xiang showed his skillfulness and his prudent attitude when dealing with the relationship between old and new relationships. He believes in “harmonious yet different”, neither repetitive nor overly unconventional. The new building not only coexists with the original telecom building in harmony, but also creates an impressive urban place in cooperation with other buildings around the Gulou Civic Square and the square itself.

Hua Xiaoning

Dean Assistant/ Deputy Director/ Associate Professor/ Master Supervisor,
School of Architecture and Urban Planning, Nanjing University

建筑城市环境 Building site in CBD area

# Shenzhen Central Business Tower



2001 / 2003

# 深圳中央商务大厦

2000年左右是深圳特区福田中心区规划设计的高潮时期。为了借鉴发达国家城市规划的经验，深圳规划局邀请了美国的SOM建筑设计事务所来进行中央商务区地块的城市设计，同时邀请项秉仁作为顾问来联络相关的专业人士。当时的中国，在制定城市规划和实施建筑之间，鲜有加入城市设计阶段的实例，项秉仁接受了这个难得的实践机会，并最终与美国的规划师同事一起完成了这个任务。随之，他也有幸接到了这一区域中一个地块的建筑设计委托——这个项目后来被冠名为深圳中央商务大厦。既作为整个区域城市设计的参与者，又作为具体地块的建筑设计师，项秉仁致力于在遵守所有城市设计导则方面作出表率，同时也希望在处理好所有建筑设计问题的前提下最大限度地满足开发商的利益诉求。

现在来看，中央商务大厦的建筑设计趋于平实，但从城市角度而言，正是这种以服从整体城市设计为建筑设计出发点的指导思想，支持了深圳市规划局建设中心区一流城市环境的设想。三层的裙房、40米高处的街墙控制线、100米以内的建筑总高等措施让建筑融入城市之中。建筑平面的排布采用最高效的方整布局形式来充分利用有效面积以满足基地地段的商业价值。框筒结构的标准层设计采用无柱办公空间，不仅可以随意划分办公空间，还可以将使用空间最大化利用。竖向办公层层高设计为3.9米，预先为智能化架空地板层的安装留有余地。

在城市设计导则中，除了对建筑的高度、建筑体块的退让规定之外，对建筑立面的线条和开窗比例也有明确的指导和约定。可能是SOM受到芝加哥规划的影响，所有的建筑都要求呈现狭窄的竖向线条，但其实在深圳，过度的窄窗并不适合当地的气候条件。项秉仁在设计过程中试图突破竖窗的限制，追求建筑内部的舒适度。然而同深圳城市规划部门协商后，最终还是选择以服从整个设计导则为原则，与相邻建筑竖向线条保持协调，通过注重使用现代材料和设计概念来表达时代特色。建筑的四个角进行敞开，以利于提供较好的室内景观，也使建筑在夜间能达到特别的灯光效果。在满足竖向线条和开窗比例要求的前提下，立面故意做了一点不对称的处理。虽然建筑单体不那么具备强烈的“视觉冲击”，但竖向线条的精心处理使整个建筑获得了良好的立面比例，具有一种庄重、高贵且现代的气质，最终的建成建筑对于城市整体形象的和谐起到了积极的作用。

实景照片 Building exterior

Around the year 2000 the planning and design of central Futian District at Shenzhen Special Economic Zone climaxed. To learn from the experience of developed countries, the Urban Planning, Land & Resources Commission of Shenzhen Municipality invited SOM, an American architectural design office, to conduct an urban design survey of the future CBD (Central Business District) and invited Xiang Bingren as a consultant to liaise with various professionals. At that time, it was unusual to have an urban design phase between the making and implementation of urban planning, but Xiang Bingren accepted the invitation and accomplished the task excellently in collaboration with his new urban planning colleagues from the US. Later, he was commissioned with an architectural design within this planned area, which was then named as the Shenzhen Central Business Tower. Here, Xiang served as both a participant of planning as well as the leading architect responsible for a specific site. Xiang Bingren decided to set an example by following all the urban design guidelines produced for this area and solving architectural issues with due diligence while meeting the interests and needs of the developer.

Though seemingly ordinary in appearance, the design of the Shenzhen Central Business Tower aligns exceptionally well with the bigger picture of Shenzhen city to create a world-class built-environment by following the urban design guidelines. Many of the building's design features such as the three-story podium, 40-meter frontage height as well as the 100-meter total height were all undertaken to help the building blend harmoniously into the existing city-scape. To respond to the high land price and business value of that neighborhood, a rectangular layout was made to maximize land utilization. Meanwhile, the floor plan uses a framed-tube structure with an 8-meter column spacing at the outer frame to enhance flexibility. As for the office layout, the column-free structure provides nearly infinite possibilities of subdivision to maximize the usable area. Moreover, the 3.9meter story height leaves enough room for the installation of smart raised flooring.

Apart from height and volume regulations, urban design guidelines specify proportional restrictions for openings and lines on a facade. Probably under the influence of the influential "Chicago Style", SOM had promoted the tendency for all buildings to be emphasized verticality, a policy at odds with the existing visual layout of Shenzhen. To acquire a more comfortable interior, however, Xiang Bingren argued that an exception be made. The planning bureau insisted that guidelines be followed to coordinate this new project with adjacent buildings. Hence the design managed to maintain an expression of the time through adopting modern materials and design concepts. There are no columns on the four corners of the building, providing an unobstructed view during daytime and a better lighting effect in the evening. While conforming to the facade design guidelines, a slightly asymmetrical touch occurred. The building exterior is not exactly very striking, however, carefully designed vertical lines present visual slenderness when the building height is limited, emphasizing a sense of solidness, elegance and modernity, and contributing to the harmonious built-environment of Shenzhen City.

# 评论
# Review

中央商务大厦的设计基于SOM的城市设计，其房地产的属性要求建筑师既要遵循上位规划的严格控制，同时也要利于销售。这也是项先生一直强调的，好的建筑师需要具备的是综合解决问题的素质，但同时又要有超前的眼光与强大的韧劲，在服务各方的前提下体现自己的建筑修为。

大厦与周边建筑既融合又具有自身独特的气质，室内的办公空间也因为其高效的平面布局深受业主青睐，这个早年的办公案例也为日后秉仁事务所在设计中更多站在城市设计的角度来思考解读项目具有一定的意义。

秦戈今

知家（Z+）空间设计主持建筑师

While the design of the Shenzhen Central Business Tower was developed based on the urban design plans of SOM, some alterations were required in order to facilitate sales. Mr. Xiang would have agreed with these priorities. He was always insisted that good architects need to have the ability to solve problems comprehensively, a forward-looking vision and the strong tenacity to maintain their ideal architectural practices under the premise of serving the interest of all parties.

On the urban level, the Central Business Center blends in with the surrounding buildings harmoniously while maintaining its own unique character. On an architectural level, the highly efficient office layout is also celebrated by the property owners. This early office project also provides certain insight for interpreting architectural projects from an urban design point of view in the future.

Qin Gejin

Chief Architect, Z+ Spatial Design

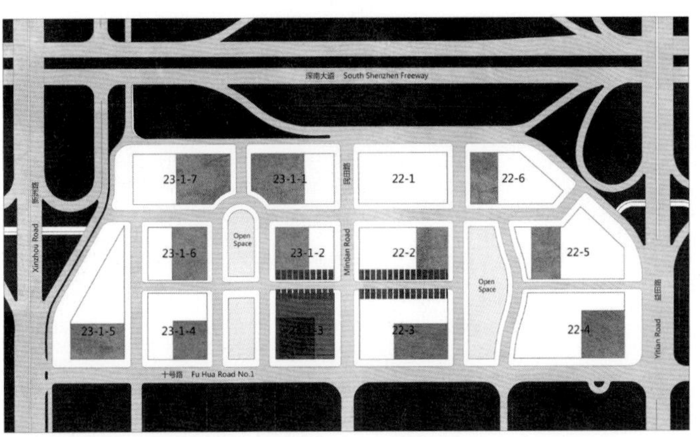


环境分析 Environment analysis

建筑沿街实景 Exterior view

# Jiangsu Mobile Communication Complex



2002 / 2006

# 江苏移动通信枢纽工程

江苏移动通信枢纽工程是一座现代化的调度、生产与办公中心，项目基地一边紧挨南京城古城墙及保护带，宁静而葱茏；另一边则是斜穿而过的城市高架路，快速且喧嚣。项目综合了生产机房和指挥调度中心、停车库、食堂和设备机房等功能，并与北侧已建办公楼协同运作。可以说，移动通信大楼是一个坐落在不太普通的用地上的非常普通的项目，而核心设计手法体现出项秉仁这一时期设计思想的成熟。设计方案干净利落，在两个层面上不乏其智慧：对于建筑项目的城市维度的思考以及对于当下现代建筑技术的拥抱。

在城市设计层面上，建筑布局可谓精致、犀利而富有戏剧性。在这块不大的用地上，移动通信大楼的设计首先面临的是北侧的喧哗主路与南侧宁静的古城墙间戏剧化的界面差异。通过对基地的解读，十二层的板式塔楼调度中心布置在北侧呈南北向，向南可远眺古城墙。为了呼应城市界面，板式办公转折至东北向贴向道路界面，形成一组锐角三角形的办公建筑平面。同时，多层的生产技术用房在板楼南侧和西侧谦逊地缓缓铺开并向城墙界面致敬。由此塑造出一南一北两个界面。在场地的规划中，北侧结合已有建筑和新建筑围合出一个高效便捷的入口广场，南侧则结合城墙景观带形成一个安静优雅的内院。两者之间通过二层通高、南北贯通的门厅进行联系，化解基地的动与静的矛盾。

为了使这种布局更好地在人性尺度发挥作用，项秉仁将办公调度大楼的一二层建筑平面架空，形成的灰空间在北侧界面结合办公入口，在南侧则结合庭院空间。这种灰空间一方面为步行环境提供了更适合的尺度，同时也为建筑体量找到了一种适宜的尺度划分，形成了下部架空段和上部主楼两个层次。事实证明，这种设计手法现在已经越来越多地在国内高级写字楼或城市综合体的设计中得以体现，它是对高密度城市中心区高贴线率建筑界面的可贵救赎。

出于对中国移动企业身份的认同及与业主的深入讨论，整个方案采用玻璃幕墙、金属幕墙以及金属遮阳系统相结合的具有时代气息的立面设计。调度大楼的北向立面主要采用明框玻璃幕墙结合狭长的开启扇，层间梁的背衬铝板玻璃层次帮助整面幕墙以横向线条含蓄地铺开。南向立面则结合了框架式遮阳百叶系统，使立面变得细腻并在光影反射下显得丰富，同时减少了能耗并提升了室内光环境的舒适性。在塔楼建筑的转角处，幕墙被完全断开，形成了具有犀利转角的两片。这种细部处理一直延续到建筑首层的架空区域，并与金属雨棚有机结合起来。建筑的裙房设计更为直接，铝板和大面积玻璃幕墙对比的手法，与主楼一脉相承，呈现了轻盈自由的立面表情。

一层平面图 First floor plan

1 办公入口
2 大堂
3 支撑网 BOSS 机房
4 主机托管机房
5 网管中心机房
6 机房办公
7 进厅
8 健身
9 休息厅
10 非机动车车库入口
11 消防车道
12 地下车库入口

The Jiangsu Mobile Communication Complex is a modern center for dispatching, production and office space. One side of the site is adjacent to the ancient Nanjing city wall with a quiet and green buffer. The other side of the site faces a noisy and fast-paced viaduct. The project encompasses production engine rooms, a command and dispatch center, parking garages, a canteen and equipment rooms, while also cooperating with the office building on the north side. To some extent, the mobile communication complex is a very ordinary project located on a special site. However, the design strategy of this project reflects the later maturity of Xiang Bingren's design philosophy. The design is neat and tidy, and integrates the architectural project in an urban context and the embracing of contemporary architectural technology.

From an urban design perspective, the overall layout of the project is delicate, yet sharp and dramatic. There is a stark contrast between the noisy urban artery on the north side and the tranquil ancient city wall on the other side. Such a disparity raised quite a challenge for the design of such a small piece of land. Through the interpretation of the site, the twelve-story dispatch center tower is arranged on the north side facing

# Jiangsu Mobile Communication Complex

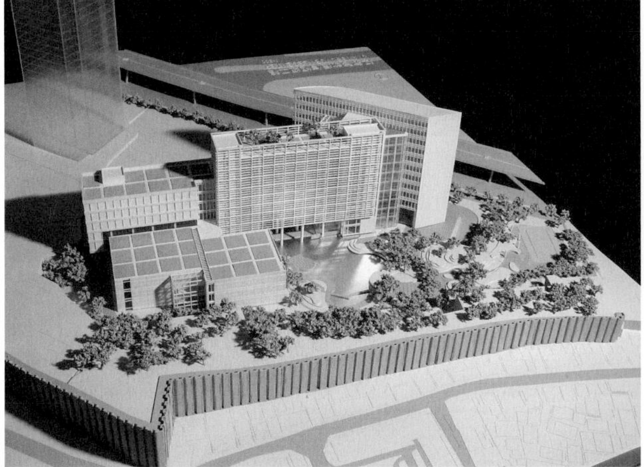

模型照片 Model photos

从古城墙看建筑 View from the adjacent ancient city wall

south, overlooking the ancient city wall. In response to the city's interface, the slab office tower turns northeast toward the road, forming an acute triangular office plan. At the same time, a multi-story production technology structure is modestly spread on the south and west sides of the slab and pays homage to the city wall , creating two different urban interfaces on the south and north sides of the site. In the planning of the site, an efficient and convenient entrance plaza is formed on the north side of the site in combination with existing buildings and new buildings. On the south side, a quiet and elegant inner courtyard is formed in collaboration with the urban landscape. A two-story hall connecting the north and south end intermediate between tranquil atmosphere and various activities.

In order to make this layout better adapted to the users' requirements and standards, Xiang Bingren opened up the first and second floor of the office dispatch building and formed a gray space with the office entrance on the north and a courtyard to the south. This gray space not only creates a suitable space for walking, but also cleverly divides the building body into two parts: the lower open floor and the upper office floor. This method of design has now been increasingly reflected in the

建筑局部 Building details

design of high-end domestic office buildings or urban complexes, which are valuable spaces among highly restricted street frontages in densely populated urban centers.

After in-depth discussion with the client and considering China Mobile's image, the entire project adopted a contemporary design combining a glass curtain wall with a metallic curtain wall and shading system. The north facade of the dispatching building mainly uses bright-framed glass curtain walls combined with long, openable windows. The backlit aluminum glass layer of the interlayer beams emphasizes the strong horizontality of the facade. A framed sun visor system helps to soften the southern facade while creating an extremely rich shading effect, reducing energy consumption and enhancing indoor light comfort. At the corner of the tower, the curtain wall was severed completely, displaying its two acute edges. This detail has been redeployed in the design of the open area on the ground floor of the building, combined with a metal awning. The podium uses a more straightforward design approach, paired with aluminum panels and large glass curtain walls, presenting the same light, and free facade as an expression as the main building.

# 评论
Reviews

记得一次项先生从国外旅游考察回来说到，看多了、设计多了还是觉得现代主义建筑的经典美经得住时间的考验。我想这个时期是项先生在新时代背景下建筑多元化发展倾向的情境中，对改良现代主义的认同，从功能维度、审美维度和技术维度，江苏移动通信枢纽工程都比较彻底地体现出这种认同。同时对建筑在城市中的角色也有非常理性的回应，从建筑布局、街区界面、空间渗透等方面都做出了审慎的思考。他曾开玩笑说上海陆家嘴的建筑就像是建筑动物园，的确，每栋建筑都要彰显自己的特色那何谈城市美学，建筑应该是以恰当的角色介入城市空间，融入环境。

董立军

厦门大学嘉庚学院讲师

在全球化的时代浪潮中坚持地域性创作是一个复杂的理论和实践问题，它不能简单地理解为重复当地建筑传统，更不能成为在设计中因循守旧、固步自封、拒绝开放吸收的借口。在勇敢面对当今时代要求、认真研究和分析基地环境和所建项目性质的基础上，我们应通过总体布局、环境营造、建筑设计乃至细部推敲，对建筑物所处地域和基地特征进行有针对性的考虑和回应，最终实现地域性与现代性的完美统一。[1]

韩冰

天华集团资深合伙人
上海天华执行总建筑师
建筑学博士

I remember Mr. Xiang once said, having recently returned from abroad, that the more he had seen and designed, the more he can appreciate the timeless beauty of modernist architecture. The Jiangsu Mobile Communication Complex Project thoroughly reflected Mr. Xiang's approval of modified modernism. Functionally, aesthetically, and technically it remains stable in an era when architecture itself was undergoing enormous diversity. At the same time, the design of this project delivered a very rational response regarding how we might understand the appropriate role architecture plays in the city. Careful thinking was made concerning the architectural layout, block interface, space penetration and other aspects. Mr. Xiang joked about how Lujiazui resembled a zoo of architectures, where every single building in an attempt to perform its own uniqueness, has lost the sense of a harmonious whole. When it comes to urban aesthetics, architecture should intervene in and integrate itself into urban space.

Dong Lijun

Lecturer, Xiamen University Tan Kah Kee College

Insisting on an indigenous creation in the tide of globalization is a complex issue. Both theoretically and practically, it cannot be simply understood as a repetition of the traditional local architecture, nor can it become an excuse to be conservative or closed-minded. We must bravely face the demands of the present age and carefully study and analyze the environment of the site and the nature of the projects to be built. On this basis, through a careful layout arrangement, space-shaping and architectural design, we should consider design questions according to the characteristics of the certain areas. In only this way might we finally achieve the perfect unity of regionalism and modernity.

Han Bing

Senior Partner, Tianhua Group
Executive Chief Architect, Shanghai Tianhua
Doctor of Architecture

1 韩冰：《地域性和现代性——江苏移动通信枢纽工程设计案例研究》，《城市建筑》2008 年 10 月。
Regionalism vs. Modernity: The Design of Jiangsu Mobile Complex, Urbanism and Architecture, 2008, Vol.10.

凯利达
小码王
少儿编程

# Tianli Central Business Tower, Shenzhen

2003 / 2005

# 深圳天利中央商务大厦

天利中央商务大厦位于深圳南山区，南山中心区的规划定位是融商业、文化、娱乐、办公和配套居住设施为一体的城市副中心。为了成功地开发建设这一城市新区，当地政府组织专业力量进行了深入策划，并做出详细的整体城市设计。天利中央商务广场位于该区11、12号地块，功能包括3栋高层办公塔楼，4 ～ 6层商业裙房以及3层地下配套设施，总建筑面积25万平方米，建筑高度130米。

项秉仁力求将建筑作为城市的基本组成部分，在严格遵守城市设计导则的基础上，无论功能布局、交通组织还是建筑形态，均以促进该区域的整体性、有序性为原则，创造有机的城市空间，为人们提供舒适、生动的活动场所。而作为商业开发项目，他又尽量合理安排建筑用地内商业和办公空间的位置，有机组织基地内外的交通流线，提升地块价值，塑造独特的建筑形态，与周边建筑共同营建良好的城市景观。

整体的建筑形体采用了简约现代的处理，上部的3栋塔楼强调高效的办公特质，平整硬朗的方形形体没有过多的变化处理，仅在塔楼两个较宽的侧面上做竖向划分以加强其纵向高耸感；裙房则以交错的形体关系渲染出浓厚的商业气氛，大型广告位及电子显示屏成为重要的构图元素。塔楼与裙房的不同特质，在对比中寻求统一，求得两者间的视觉对比。

仔细观察会发现，建筑材质的选择和运用有特别的考虑。塔楼的外围护玻璃幕墙可能是出于视觉效果和节能设计的双重目标，选用了四种不同质感的玻璃。银蓝色的玻璃配以深灰色铝合金框料，玻璃面随日间光线的变化与天空相映生辉。裙房的材料以石材为主，凹凸错动的形体分别配以白麻、黑洞石、中国黑、福建黑、新疆雪莲花等不同石材，塑造建筑向上的动感。其中，石材表面的斧凿、烧毛、抛光等处理更有意凸显精工细作的细节之美。灰色基底配以黑色石材与清玻璃的组合，显示出高贵内敛的气质。现在看来，虽然项目整体形体简洁，但由于立面错动的组织和细致的雕琢，仍然使得这组建筑高耸典雅且充满生机。

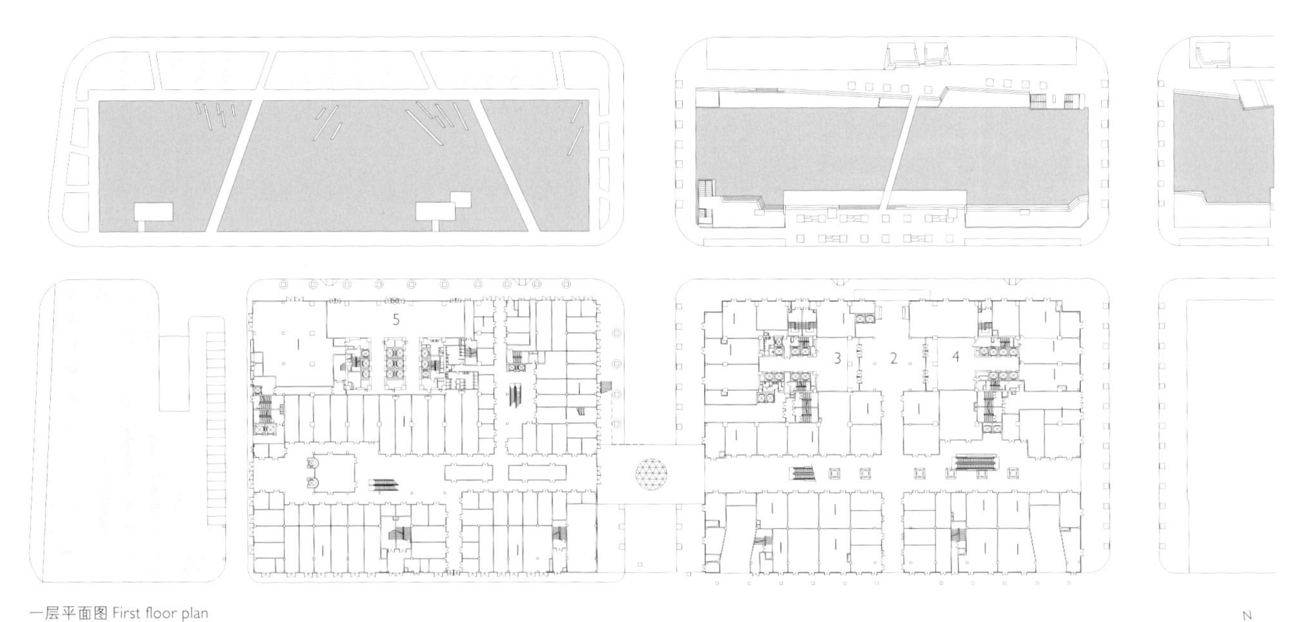


一层平面图 First floor plan

1 商铺
2 公共开放空间
3 办公楼 A 楼大堂
4 办公楼 B 楼大堂
5 一期办公大堂

The Tianli Central Business Tower in Shenzhen was commissioned as a larg high rise commercial complex in Nanshan District. Central Nanshan District was planned to be a sub-center for the city, integrating commercial, cultural, entertainment, office and supporting living facilities. To successfully develop and construct this new urban area, the local government organized professionals to carry out in-depth planning and to create a detailed and comprehensive urban design plan. Tianli Central Business Plaza is in the No. 11 and No. 12 parcels. Its functions include 3 high-rise office towers, a 4 ~ 6-story commercial podium and 3 underground supporting facilities with a total construction area of 250,000 square meters and a building height of 130m.

At the beginning of the program, Mr. Xiang sought to make architecture an essential part of the city. He strictly observed the urban design guidelines, both in terms of functional layout, for traffic organization and architectural forms. He additionally promoted the integrity and orderliness of the region to create an organic urban space, providing residents with comfortable, lively activities. For such a commercial development project, he worked to arrange the commercial and office volumes on the land as reasonable as possible, to organize the traffic inflows and outflows organically and to generate the most overall value out of the parcel. He further endeavored to shape the unique architectural image while co-operating with the neighboring buildings to create a positive, well-proportioned urban image.

The form of the building is simple yet modern, with the upper

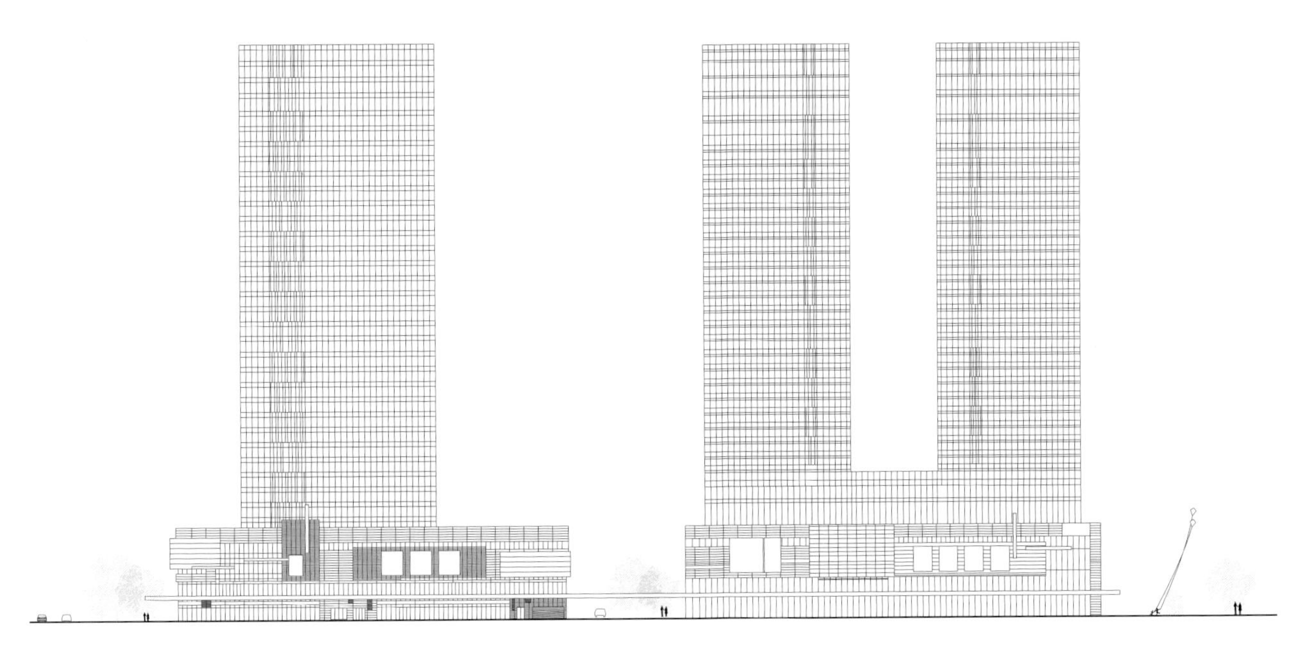

南立面图 South elevation

three towers featuring a rectangular layout that is clean and neat without excessive variation to emphasize a professional office image. Vertical divisions are only made on the two wider sides of the tower to enhance its visual verticality while the staggering volume of the podium creates a strong commercial atmosphere. The large advertising spaces and electronic displays serve as important compositional elements. Despite the different characteristics between the towers and the podium, the two designs work together in seeking a unity and visual balance overall.

If you look closely, you will find that the design is also ingenious in the selection and use of materials. Likely installed for visual and energy-saving purposes, the glass facade on the outer perimeter of the tower is composed of four different types of glass. The silver blue glass is paired with a dark gray aluminum frame. The glass's color will shift in the daytime light, glowing under a bright sky. The concave and convex podium are mainly covered ing stone, white linen, black hole stone, China black, Fujian black, Xinjiang snow lotus and other varieties which together help to create an upward visual momentum for the building. Moreover, the axing, singeing, polishing and other processing technique of stone surface helps to highlight the beauty of its luminescent and textual details. The combination of black stone on a gray base and glass reveals an elegant yet restrained temperament. Though the overall design is simple in form, the organization and meticulous care of the facade still make this building elegantly towering and full of life.

裙房效果图 View of the podium

## 评论
## Review

项目位于深圳南山商务文化中心区二层步行街西端起点处，项先生在平衡地块经济效益与城市环境效益之间做出了恰当适宜的设计决策。三栋超高层塔楼采用极简的方形体块，既实现了使用效率最大化，也取得了经典的现代主义审美特质，极为典雅大气，在高楼林立的中心区内也能独树一帜。裙房部分充分考虑了商业特性，内部空间具有较强的适应性，在造型上采用体块穿插的手法活跃商业氛围。项目与区域内毗邻地块形成融为一体的区域购物环境，成为南山区的新地标，舒适的办公及商业空间也使人们体会到无穷的都市魅力，为凝聚区域活力作出了贡献。

董立军
厦门大学嘉庚学院讲师

This building is located at the west end of the pedestrian street in Nanshan Business and Cultural Center. It is a balanced result between economic benefits of the plot and environmental benefits for the city. Three super high tower blocks adopt a simple square shape, a distinguishing modernist feature, to maximize the efficiency of use. The annex is carefully designed for business use, with highly adaptable interior spaces and interspersed blocks to activate the commercial vibe. Together with the adjacent land, this project has created an integrated localized shopping environment and has become a new landmark in Nanshan District, showcasing urban charisma and vitality to both locals and visitors.

Dong Lijun
Lecturer, Xiamen University Tan Kah Kee College

从公园看建筑 View from the adjacent park

# Jian Hui Tower Renovation, Shanghai



2006 / 2011

# 上海建汇大厦改造

地处上海市徐家汇商圈核心位置的建汇大厦建于20世纪90年代，是这座城市改革开放初期建造的高档涉外高层办公建筑之一。然而，随着徐家汇商圈日新月异的发展，这栋当年外观采用珠光面砖和银白色单层铝合金外窗的建筑已显得陈旧落伍，加之周边新颖时尚的现代高层写字楼楼宇林立，更使之相形见绌。面对城市建设的不断推进，大楼的开发商和城市规划部门都希望对这栋楼宇进行一次较大程度的更新改造，力求通过全新的形象设计，提升大楼的总体环境和品质，推进城市可持续发展的步伐。一定程度上，这个项目也可以视作一次历史建筑和城市风貌的更新设计。

在认真调研了这栋主楼35层的办公楼的建筑现状和图纸档案之后，项秉仁向业主提出更新策略，即不仅需要重点优化建筑造型和立面用材，同时兼顾大楼原有结构和设备状况，考虑改造的可操作性和成本可控性，更需要提升楼宇的节能和生态性能，以符合高档物业的时代需求。

最终的改造方案中，立面处理手法强调整体性和时尚感，并注重细节。塔楼主体部分为对称的四段圆弧，在材料上选用低反射的双层钢化Low-E中空玻璃，通过横明竖隐的线条疏密变化形成一层有质感的完整外表皮。配合原有墙体开窗洞口及通风，达到类呼吸式幕墙的效果。建筑四角选用复合铝板并加以竖向的凹凸线条来强化主体建筑的力度与挺拔感，并在顶部做内向切角，以承托上部建筑。为了进一步强调建筑的整体性，顶部机房、水箱层平面被适当加大以对应标准层的平面关系，通过外圈规则排布的装饰板片，形成顶部特有的造型与表皮肌理，优化大楼的整体比例。

由于大楼已使用多年，为了尽量减小改造施工工程对现有楼内租户正常办公的干扰，所有外窗考虑在厂家进行标准的单元式制作，并利用非工作时间进行安装。在施工时要求注意对原有墙体的保护，防止面砖脱落，同时满足保温与防水的构造要求。在进行建筑上部结构改造时应尽量保证热泵机组位置不变，同时采用钢结构进行结构外挑设计。根据造型的需求考虑加大电梯机房层、水箱层平面。在有利于整体造型效果的同时，也增加了有效的使用空间。建汇大厦在2010年完成了改造，不仅建筑物的外形和内在品质得以更新，而且也提升了徐家汇中心区域的城市面貌。

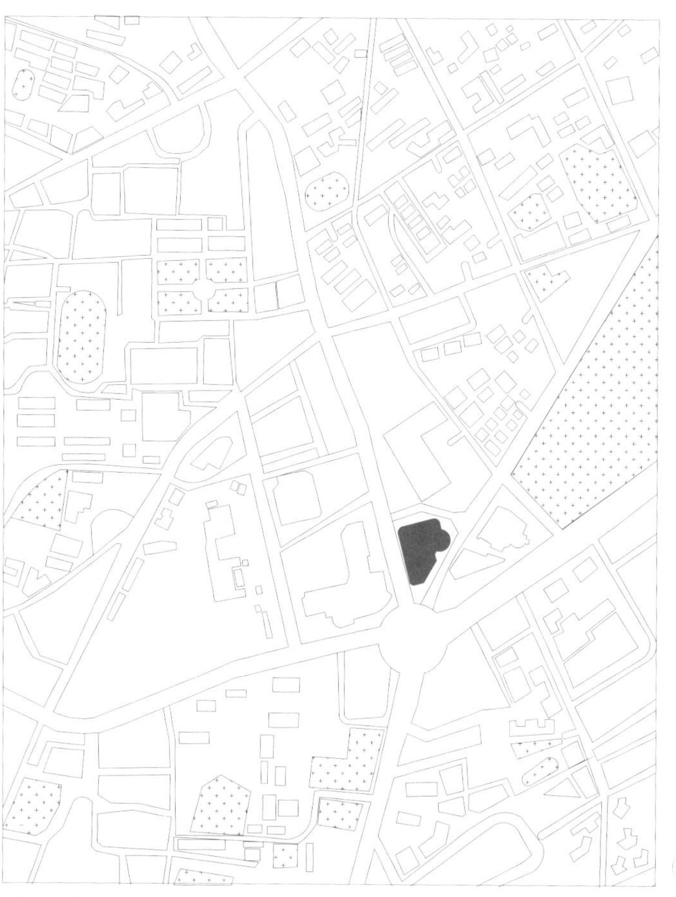

总平面图 General site plan

改造前 The building exterior before renovation

The Jian Hui Tower, located in the core of Xujiahui Business District, Shanghai, was built in the 1990s. It was one of the high-rise, high-end office buildings for foreigners built during the beginning of the reform and opening up. However, with the rapid development of the Xujiahui Business District, the appearance of this building, which adopted pearlescent face bricks and silver-white single-layer aluminum alloy exterior windows, seemed outdated. In addition, the Jian Hui Tower was overshadowed by the surrounding novel and stylish modern high-rises. As a result of the boost to urban construction, the developers and City Planning Administration wanted to give the building a makeover from a brand-new image design and a promotion to the overall environment and quality. It was hoped that such a renovation would propel the sustainable development of the city. This project might be regarded as a renovated design for historic architecture and city makeover.

After elaborately investigating the current condition and original documents of this 35-story building, Xiang Bingren proposed a renewal plan to the client: the building should not only emphasize an optimization of its architectural image and building materials but also account for the existing structures and conditions of equipment to assess the operability and control costs of the project. To meet the expectation of high-end properties, improving the energy-saving and ecological attributes of the building was essential. The facade's renovation required style, integration and attention to details. The main body of Tower was four symmetrical arcs, selecting bilayer tempered Low-E insulating glass that has a low reflection as a material. They formed a textured intact surface by altering the density of patent horizontal lines and latent ver-

# Jian Hui Tower Renovation, Shanghai

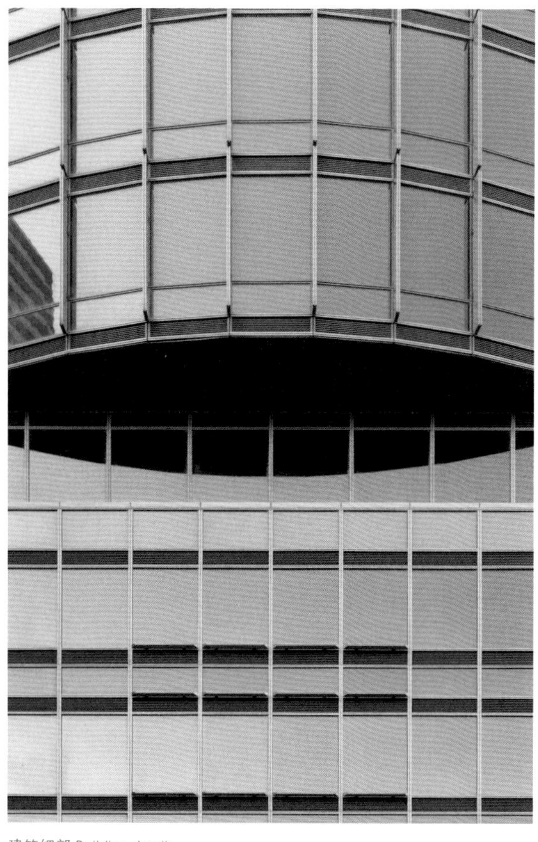

建筑细部 Building details

tical lines. With the existing holes for windows and ventilation, the main body achieved a result similar to a respiration-type curtain wall. The four corners of the chosen building are composed of aluminum plate ad combined with concave-convex lines to reinforce the dynamic and upright aesthetic of the main volume while also cutting corners inwardly on the top bore the upper part of the building. To further underline the integrity of the building, the machine room on the top and the cistern were enlarged appropriately to match the corresponding plane relation by shaping a unique style on the top and surface texture from regularly distributed decorative sheet bars on the outer ring. Such an arrangement served to optimize the overall proportion of the building.

Seeing that the building had been used for many years, in order to minimize the impact of construction on tenants, all external windows were standardized, pre-manufactured and installed during non-business hours. During construction, special attention was paid to the protection of existing walls and the prevention of face bricks falling off. Detailed requirements for heat preservation and water proof were met. When renovating the upper part of the architecture, the heat pump system remained undisturbed and steel was used to construct the exterior structure. Based on the styling requirement, the elevator machine room and the cistern were magnified, improving the overall styling as well as increasing the effective usable space. The renovation of Jian Hui Tower was finished in 2010, and it not only improved the appearance and interior quality but gave the core of Xujiahui Business District a new look.

# 评论
# Review

建汇大厦始建于1996年，项目是一个建筑外立面改造工程，它的特殊性在于施工的同时要不影响内部办公，作为环绕徐家汇的几栋核心办公塔楼之一，如何让一个90年代的老大楼既符合当代审美又体现当代技术的革新，它的形象重塑就变得尤为重要。

设计上希望达到的效果是化繁为简，梳理原有建筑体型，将原有变化的体块整合成一个整体，让新的立面挺拔干练，在顶部的处理上将建筑女儿墙做了斜切的处理，来承托由铝合金肋板包裹的圆形顶部，显得轻盈灵动。建筑底部的办公入口及建行入口，细腻的金属线条结合灯光处理与整体建筑统一并增加了仪式感。

秦戈今

知家（Z+）空间设计主持建筑师

Jian Hui Tower was built in 1996, with the objective of renovating its facade while keeping the building optional as one of the major office towers surrounding Xujiahui sub-center. The challenge was how to keep an old construction from the 1990s in accordance with contemporary aesthetics and reflect the innovation of contemporary technology.

The design aims to simplify the old appearance by integrating the overly fragmented volume into one piece, which lead to a rather sleek facade. Also, the building's parapet wall was slanted to support the dome and wrapped in aluminum alloy ribs on the rooftop to create a light and floating crown. Delicate metal linings combined with proper lighting around the entrance of the tower lobby while endowing the CCB's service lobby a dose of rituality.

Qin Gejin

Chief Architect, Z+ Spatial Design

行政办公主楼入口 Main entrance of the new government center

# Ningbo New Municipal Centre



2006 / 2012

# 宁波市东部新城行政中心

跨入21世纪，宁波市政府开始雄心勃勃地筹划宁波城市发展的战略蓝图。东部新城，此后十几年间城市新中心的迅速崛起由此拉开了序幕。2006年，作为东部新城的先导项目，行政中心的国际设计竞赛启动。

项秉仁对参与行政中心的规划设计竞赛充满了期待。多年国内外的生活经历使他对于不同社会政治文化背景下的政府建筑有较直观的认识。他曾在美国旧金山市政府大楼的广场上看到无家可归的流浪者随地蜷缩，也注意到在东京市政府大厦的首层和二层都是对公众自由开放，并陈列着这个城市的展示图片，充溢着亲和力。而与之相比国内的机关大楼则戒备森严。他充满激情地希望将自己对行政空间的全新理解灌注于行政中心的设计中。新的行政中心建筑该如何体现宁波这座历史悠久、文化底蕴深厚、经济发达的东部沿海城市在今天的特殊魅力？新的政府大楼形象能否打破国内大多数城市庄重有余而亲和不足的刻板印象，以平和亲民的形象示人？

行政中心的规划尊重以“行政湖”为中心的构思理念，延续南北向城市轴、东西向生态走廊，保持“一心双轴”的连贯与完整。行政办公楼布置于南北轴线的正中，统筹全局，构建以办公主楼为核心的功能建筑，其他功能建筑（会议中心、武警信访等）为辅楼的建筑群体形象，塑造城市轴区域景观序列的标志性起点。行政主楼的设计一扫以往严谨对称的设计惯性，将市人大、市委、市政府和市政协四套班子，由东到西，以不对称的构图形成均衡完整的建筑形象，具有张力又不失稳重，富有时代感。与此同时，四个庭院面向城市广场，建筑空间与城市空间内外融合、连成一片，形成通透而开放的空间，由内而外展现了市政府民主、亲民的姿态。主楼正中心的市民活动大厅，呈现玲珑通透之态，便于举行各种仪式以及市民集会活动，展现政府“以民为本”的形象。

开放性的行政中心造就了一个既能满足物质和精神功能要求，又充满活力和民主的精神场所。建成后的行政主楼以适度的体量和大胆的、打破常规的不对称立面设计博取了公众的注意，其位于中部的透明玻璃大中庭可供市民参观浏览，也提供了一处展示宁波历史文化和今日成就的公共空间。行政中心设计的脱颖而出让我们再一次看到了项秉仁身上所富有的创新精神和理想主义情怀。正是建筑师对于公开、民主精神理想的大胆追求，成就了宁波政府新时代下崭新的政府形象。

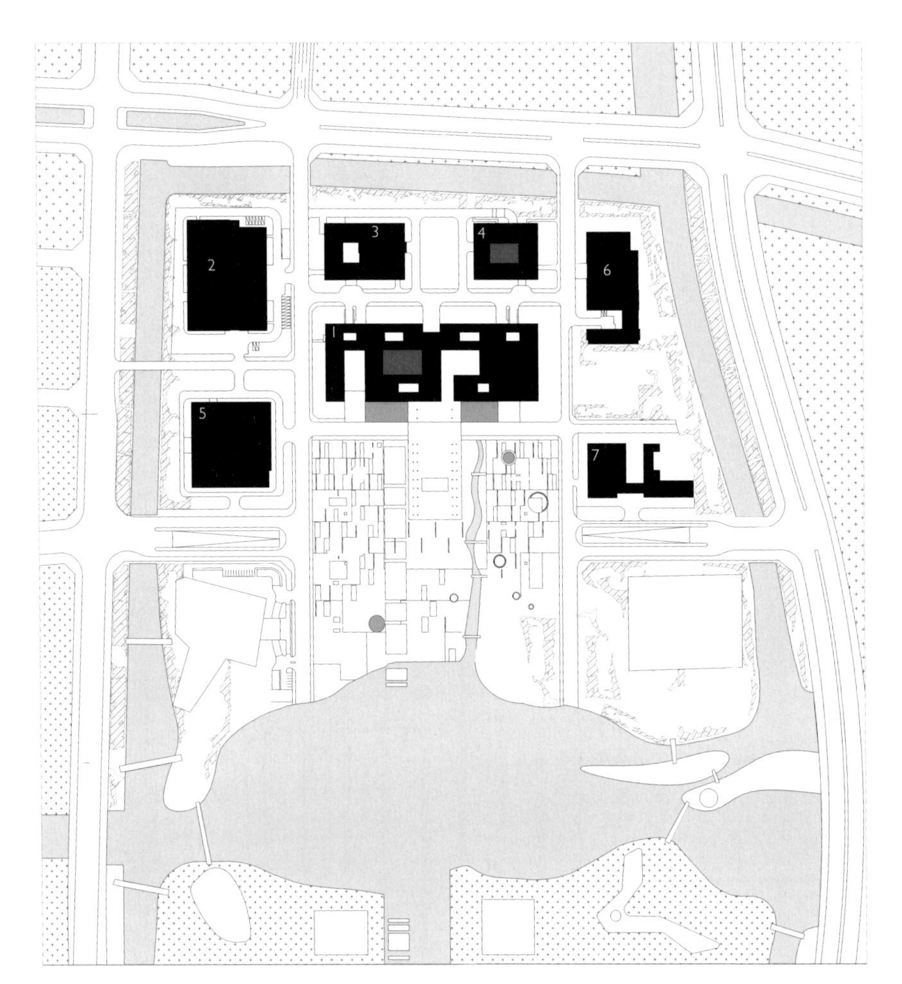


总平面图 Site plan

1 行政办公主楼
2 会议接待中心
3 综合商务楼
4 信息商务楼
5 行政服务中心
6 后勤服务中心
7 信访武警中心

Around the beginning of the 21 century, the Municipal Government of Ningbo City began planning for a strategic blueprint for urban development. In the past decade, Eastern New Town had developed rapidly into the new city center of Ningbo. The international design competition for the New Municipal Centre was launched in 2006 as a pilot project.

It was extraordinarily exciting for Xiang Bingren to participate in the planning and design of the New Municipal Centre. With his previous experiences of living in China and abroad, Xiang had developed an intuitive understanding of government buildings under the influences of varying social, political and cultural backgrounds. Xiang recalls having once walked past the San Francis City Hall in the United States where he witnessed homeless people freely sheltered out front without any disturbance from passers-by. Another time, he noticed that the first and second floors of the Tokyo Metropolitan Government Building were open to the public, with a free photograph exhibition of the city, representing the democracy and approachability of government. In contrast, the government buildings in China are always heavily guarded

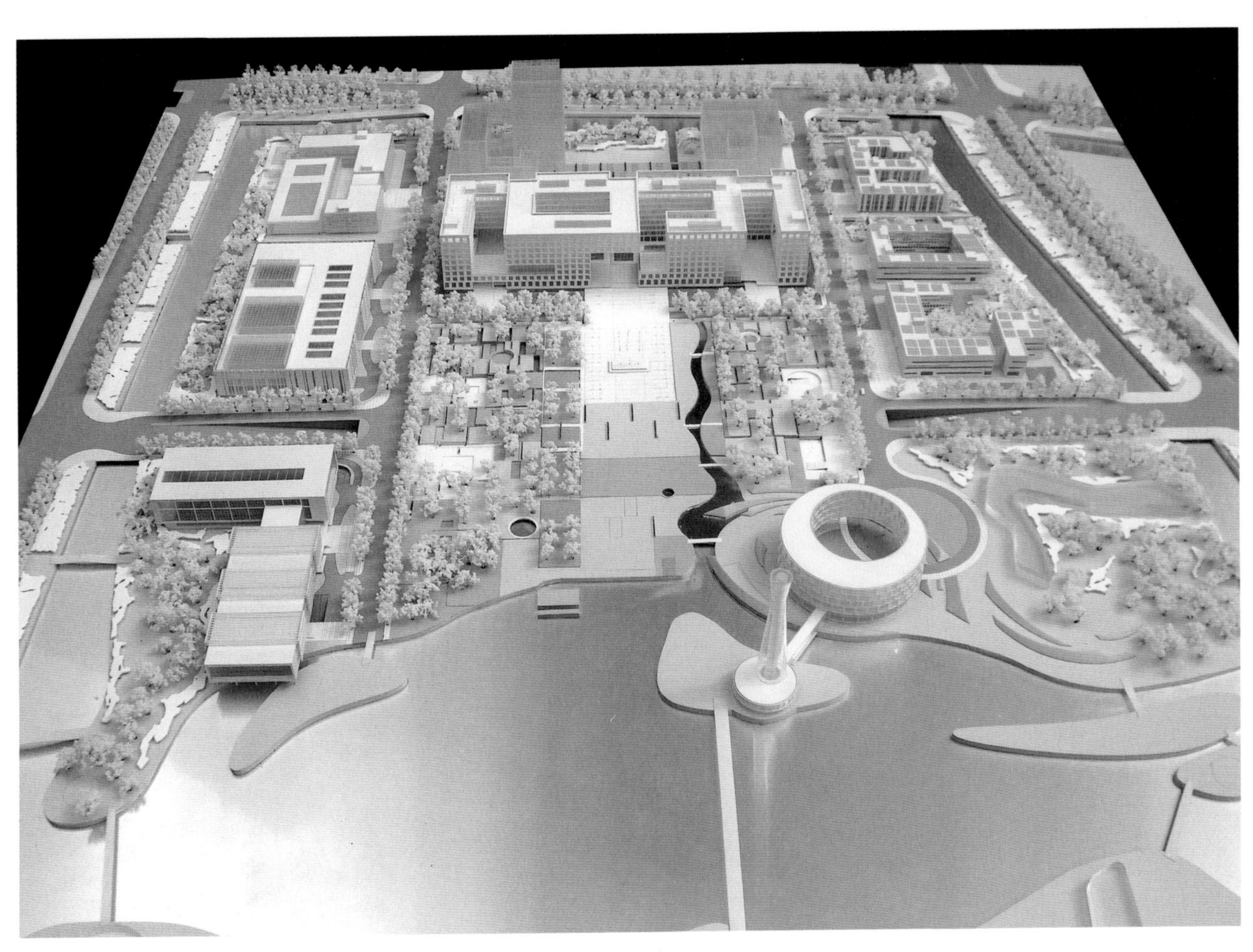

模型 Model

by security lacking any sense of cordiality or approachability. Xiang was passionate about implanting his personal understanding of governmental administrative spaces while representing his vision of democratic society in his design of the New Municipal Centre. How should the new government building display the charm of Ningbo, its profound history and deep cultural connotation? Could the new Municipal Centre break the stereotypes of government buildings being solemn and exclusively rooted in traditional forms?

The construction plan of the New Municipal Centre is centered around the "Municipal Lake", extending along the north-south axis of Ningbo City and the east-west axis to the Eco-Corridor. This particular configuration represents the continuity & integrity of "one center, two axes". The main executive office building resides in the middle of the north-south axis as the core functional structure along with other ancillary buildings, such as the conference center and guardhouse. Different from ordinary the symmetrical designs for government buildings, the main executive office building possesses a unique, architectural image full of strength, dignity, and contemporaneity balanced by the asymmetrical

正立面实景 Front view of the center

# Ningbo New Municipal Centre

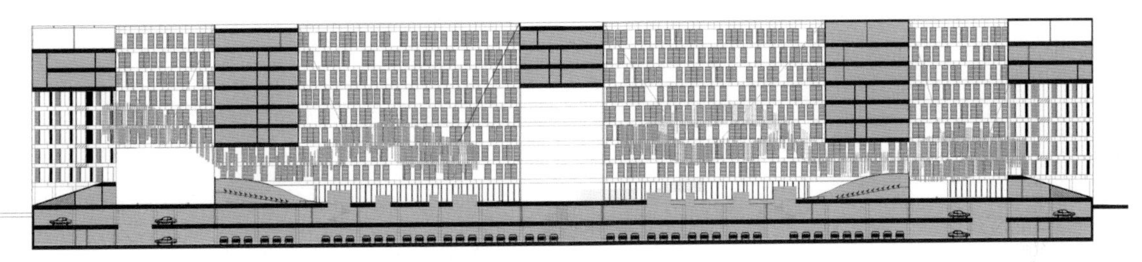

贯通的中庭空间

The interconnected atrium space

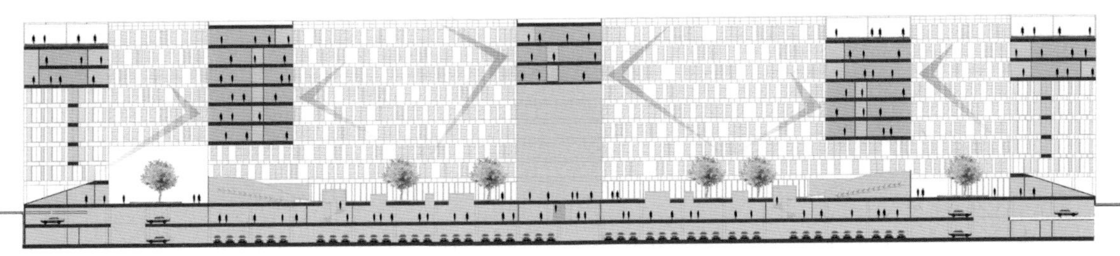

东西向房间视线与景观

East-west room views

composition flowing from east to west. Also, there are four courtyards facing the city square which integrate the architectural and urban spaces while creating a transparent open space reflecting the municipal government's democratic values and approachability to the public. The recreational lobby at the center of the main executive building allows people to organize a variety of activities and gatherings, representing the "people-centered" policy of the Ningbo Municipal Government.

The New Municipal Centre provides an open space full of energy and democracy to meet people's material and aesthetic needs. After construction, the main executive building has attracted the public's attention with its comfort and creative asymmetrical facade design. The spacious atrium with its transparent windows highlights an exhibition of Ningbo's rich cultural and historical achievements and is available for the public to visit. The outstanding design of New Municipal Centre explicitly reflects the creativity and idealism of Xiang. Because of the designer's bold pursuit of democratic values, the Ningbo Municipal Government can re-fashion a brand-new image for a new era.

建筑内庭 Courtyard view

# 评论
# Reviews

公共性和秩序感是宁波行政中心有别于同时期大型政府行政中心的最大特点：在建筑功能上用简单模数解决四个职能部门的复杂功能问题，通过分离单一建筑形式语汇形成不对称的动态平衡，从外部市政广场到内部公共空间在视觉上的严谨对位联系，以及中间挑空公共中庭的独特处理。公共性和秩序感在这个独特的行政建筑中达到了完美的融合。

吴欣

天华集团副总建筑师
上海天华执行总建筑师
建筑学博士

A sense of publicity and order differentiate Ningbo Municipal Centre from other contemporaneous large-scale governmental buildings. Functionally speaking, the design fulfills the many utility needs of the four departments with simple modularity. Formally speaking, the single architectural design achieves an asymmetrical and dynamic balance through separated volumes. Moreover, the rigorous visual alignment of the municipal plaza, internal public spaces, and the unique design of the atrium add a sense of publicity and order to the uniqueness of this administrative building.

Wu Xin

Vice Chief Architect, Tianhua Group
Executive Chief Architect, Shanghai Tianhua
Doctor of Architecture

# Ningbo Housing & Urban-Rural Committee Building

2008 / 2011

# 宁波城乡建委大楼

随着宁波东部新城总体开发建设的推进与深入，一大批政府职能部门跟随着行政中心的迁址而迁往新城区，宁波城乡建委大楼项目正是其中之一。凭借着宁波东部新城行政中心的成功，项秉仁再次通过设计竞赛获得了城乡建委大楼的设计委托。项目基地位于东部新城中央走廊南侧的混合使用区，四边临路且平坦方正，整个片区在设计伊始仍是一片不毛之地。这块看似平淡无奇的地块实际上却需要从总体规划上考虑多重环境因素的影响：用地北侧的中山路是宁波城市发展的主轴，具有强烈的都市生活属性；西侧为规划公共绿地，东北侧为规划市政湖公园，都侧重于柔性绿色的景观属性。从城市设计的视角综合考虑功能、形象、景观、环境、交通等因素，以前瞻的眼光对周边环境做出回应，是对建筑师综合理解城市属性的考校。

整个大楼形体由L形的主楼和裙房衔接成U形布局。L形的主楼沿西北两侧布置，响应城市发展规划的都市轴属性并且最大限度地吸纳周边景观优势。裙房的U形体量围合形成南向入口广场，提供了重要的人流集散场所和休息交流场所。它既联系了河清路、规划路与基地周边的沿河景观，创造了独具特色的城市公共空间，也使整座建筑形成统一的整体。通过对外部空间，广场和建筑的界面、形体的处理，整体建筑与周围城市环境在城市维度上产生和谐的相互关系。

在造型设计上，建委大楼以“砌筑城市”为设计主题，将石材和玻璃运用“凹凸、错叠、重复”等方式，创造出一种独特的“砌筑感”，并通过对“砌筑感”的表达来隐喻建委“构筑城市”的精神内涵。这一主题既与建委作为城市建设者的重要职能相契合，也体现了建委与其他办公机构相区别的独特的建筑个性。

漫步建筑内外，我们可以感受到变化丰富、浑然一体的建筑外部空间，以及室内外环境有机交融的空间体验。建筑外部空间对城市空间开放，提供了更多的交流场所；建筑内部则通过各个空中花园、屋顶绿化、挑空中庭为办公空间带来流动性。整个建筑呈现出国内行政建筑少有的开放、透明和清新，这是政府机构廉政亲民的表现，也是建筑师价值取向的表达。

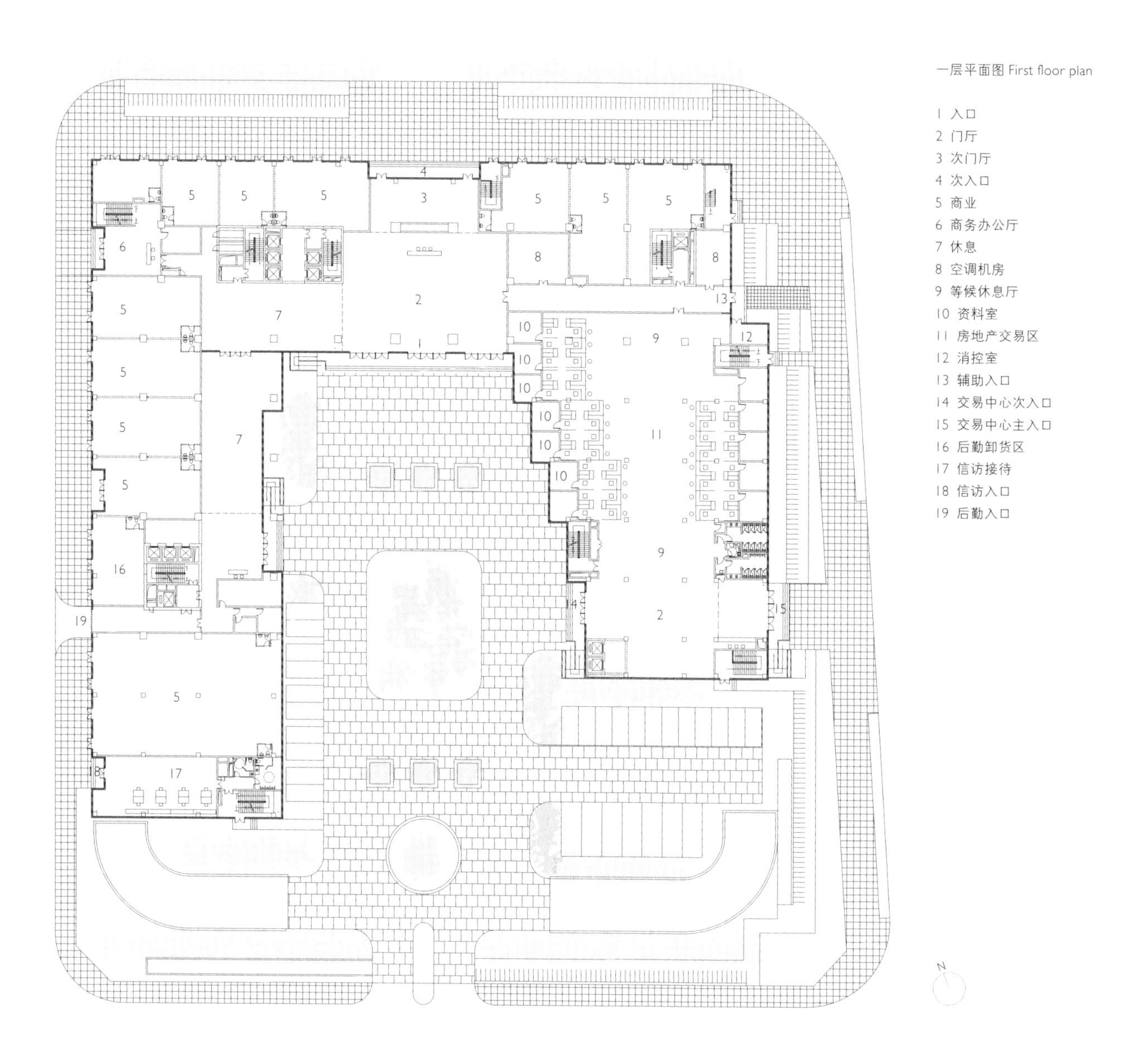


一层平面图 First floor plan

With the development of Ningbo New Town, a large number of government buildings, together with the Municipal Centre relocated there. The Ningbo Housing and Urban-Rural Committee Building was designed and built to accommodate this relocation. Due to the successful design of the Ningbo New Municipal Centre, Xiang Bingren won the design competition for the Housing and Urban-Rural Committee Building. The project is located in a mix-use area to the south of the central corridor of the new town. Four sides of the site are flat and adjacent to the road. The whole area was under-developed at that time. This seemingly mundane plot required thorough consideration of impact from multiple environmental factors for its overall planning. Zhongshan Road, to the north of the site, is the principal axis of urban development and has a strong urban character. The west is planned to construct public green space while the northeast was offered as the site for the Municipal Lake Park. Both of these projects focused on scenery attributes. The site required the architect's comprehensive understanding of city development, i.e. consideration of functionality, urban image, landscape, environmental impact, transportation from urban design perspective. The architects worked to design accordingly.

The whole building is composed of an "L" shaped main tower and a "U" shaped podium. The main building is along the northwest side, responding to the urban development axis and maximizing the advantages of surrounding landscape. The south entrance square, which is enclosed by a

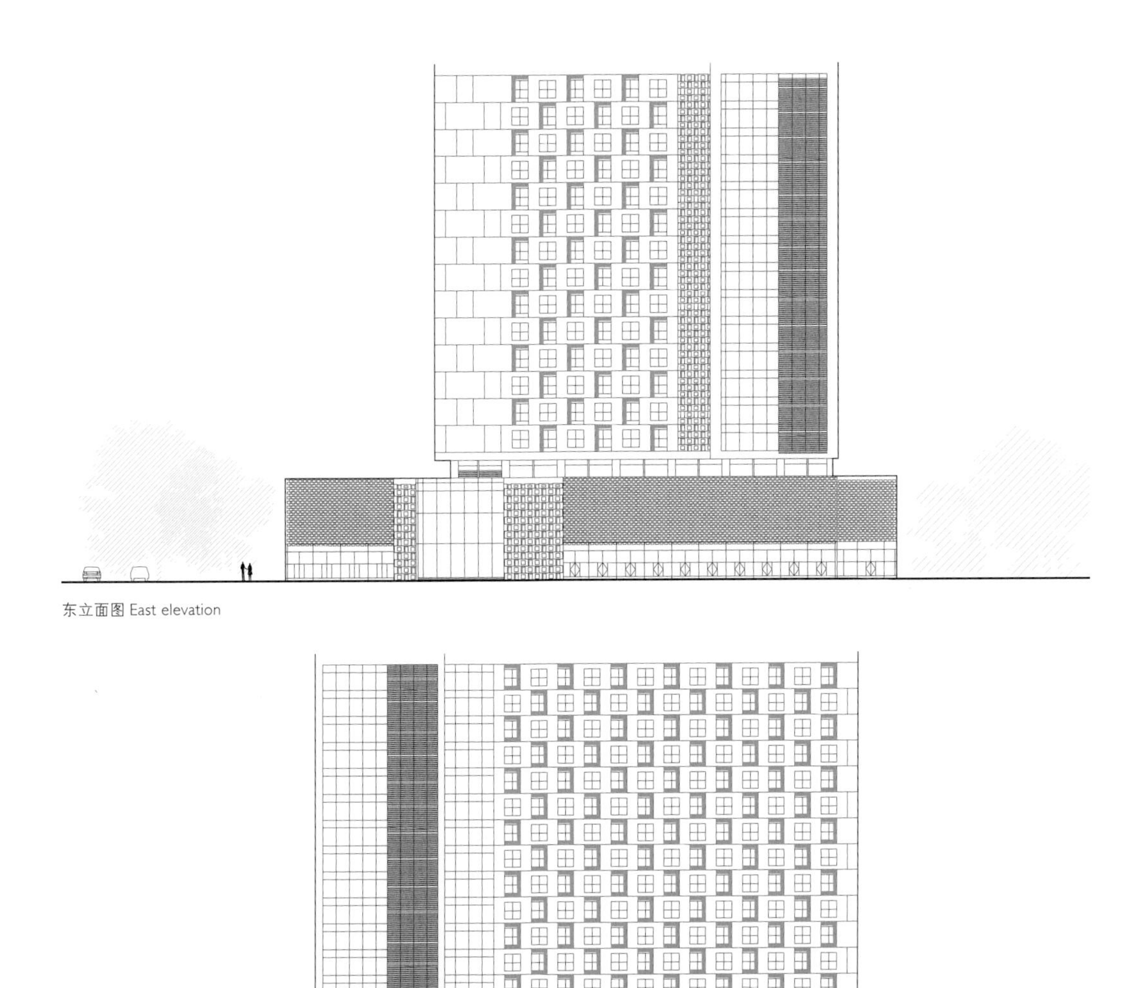

东立面图 East elevation

南立面图 South elevation

U-shaped annex, provides an important space for people to gather and rest. It connects to Heqing Road, Guihua Road and the surrounding landscape, creating a unique urban public space while unifying the architectural complex. Through the external space, square and the building interface and shape, the overall architecture integrates with the surroundings harmoniously.

The form of Housing and Urban-Rural Committee Building originated from the theme of "City Masonry" while creating a sense of "urban handcraftsmanship" through the concave-convex tension, repetition, and staggering of stone and glass. The design theme is not only in line with the Committee's important role as the city maker, but also distinguish es the buildings from surrounding office towers where other government departments are located.

Walking in and out of the building, visitors are treated to a spatial experience with the changeable and unified exterior space along with the perfect blending of indoor and outdoor environments. The outer space opens to the public, providing more communication places. The interior of the building provides flexibility for the office space and the infusion of indoor and outdoor space by including various hanging gardens, roof greener and void structures. The building presents an open and transparent atmosphere, which is rarely seen in domestic administrative building design, showing the incorruptibility of the government and closeness to its people in addition to the value of the architect.

南
琼
林
路

# Ningbo Housing & Urban-Rural Committee Building

沿街面实景 View along the street

砌筑感的建筑细部 Exterior details

# 评论
# Reviews

项秉仁先生极善运用简单的几何模式，营造结构清晰的系统和高水准的建筑语境。他常将立方体和格子融入现代时尚之中，然后将其转换为一种清新、纯粹和深刻的建筑形象，建委大楼也是如此。整个作品简洁、理性、洗练却不显单薄，当阳光沿着建筑倾泻而下时，浓烈的阴影与清新的浅色石材形成强烈的对比，展现出富有张力的建筑美。

颜莺
上海秉仁建筑师事务所合伙人、副总建筑师

建委大楼的成功现在看来也部分地因为时代机遇。一是使用方的角色令相关各方格外认真努力；二是代建方充满了夺取鲁班奖的决心；三是当时宁波的政治环境民主且开放，给予建筑师更多的话语权。整个建筑设计几乎都是按照建筑师坚持的想法进行的：大厅梁抬柱架起了上部所有楼层；入口玻璃幕墙与6米悬挑雨篷采用轻盈的无立柱自承重设计；彩釉图案根据特制的1:1图形进行定制打样；幕墙每个细节都以最佳方式得以刻画；所有的装饰材料全部由设计团队选定并进行等比例样板墙比选……

这些有利条件的出现，使得设计团队更有动力，项老师也为此投入了更多的精力与时间。他时不时会跟我说他对于细节的想法，透露出他对于设计的持续思考。建委大楼立面传达的是砖石堆叠的意象，仅关于体块之间的宽缝尺寸这一细节，我们就进行了多次的探讨与方案比选，最终以40毫米定稿。

我们的努力与真诚使合作方与代建方有了更多的信心。幕墙设计单位每周定期来上海与我们讨论从构造到材料的一切细节；代建方也更多地信任项老师的专业判断，打消了各种疑虑。而各方的肯定与积极反馈又增强了我们的信心和动力，良性的互动促成了超高的设计还原度和设计品质的提升。

史晨鸣
上海林同炎李国豪土建工程咨询有限公司 ANT 建筑创作与研发中心主任、主持建筑师

Mr. Xiang Bingren is extremely talented at using simple geometric patterns to create a clear structural system and high-level architectural context. He often incorporates cubes and plaids into modern fashion and then transforms it to a fresh, pure and deep architectural image. This is what he does with the Housing and Urban-Rural Committee Building. The design is concise, rational, and sophisticated but never weak. As sunlight pouring down the building, the contrast of dark shadow and bright stone presents an architectural beauty full of engaging tension.

Yan Ying
Partner/ Deputy Chief Architect, DDB Architects Shanghai

In retrospect, the success of this building can be attributed to several factors. First of all, the role of the client in itself compelled all parties involved to make extra efforts. Second, the contractor was determined to win the Luban Prize. Third, the relatively open political environment in Ningbo at that time empowered the architect with such control that nearly all of the design ideas were realized. For instance, use beam- columns in the hall to support all the upper floors; adoption of the entrance glass curtain wall and the free-standing 6-meter cantilevered awning; the colored glaze pattern that was customized based on a 1:1 ratio figure; every detail of the curtain wall was presented in the best way; all the decorative materials were hand-picked by our design team and then compared with each other using proportional sample walls.

These favorable conditions definitely motivated our team to invest lots of energy and time into the project. Mr. Xiang would share with me his thoughts on details, on the whole design from time to time. We worked carefully on every detail. For a simple question like how wide the gap between the blocks should be if we want the facade to convey the imagery of stacking bricks and stones, we went through rounds of discussions and scheme comparisons before finally settled on 40mm.

Our efforts gained us more confidence from our partners and contractors. The design firm for the curtain wall for example came to Shanghai regularly every week to discuss with us all the details from structure to materials; our contractors trusted Mr. Xiang with all sorts of questions. Their affirmation and positive feedback in return encouraged us. It's this kind of healthy interaction and collaboration that can lead to high performance and high degree of realization.

Shi Chenming
Director/ Chief Architect,
ANT Architectural Creation R&D Center, Lin Tung-Yen & Li Guo-Hao Consultants Shanghai Co., Ltd.

北
青林湾街

# Ningbo Electric Power Authority Building

2008 / 2012

# 宁波电力大楼

在宁波东部新城的一系列行政办公建筑的成功实践，为项秉仁在宁波的公共建筑设计领域赢得了一致的赞誉。他所倡导的民主、透明、开放、亲民等特质，与理性典雅的建筑形象相结合，颠覆了人们对传统行政办公建筑的刻板印象。正当其时，宁波市电业局开始筹划在海曙区兴建一座新的办公及技术大楼，他们迫切希望消弭公众对于“电老虎”的负面认知，而项秉仁所展示出的建筑理念与此诉求不谋而合。

基地呈不规则的长方形，西北两侧是城市道路，南侧是一块商务办公用地，东侧与河道之间是30米宽的景观带，各个界面的属性非常明确。在解读了设计任务之后，项秉仁将业主的功能诉求归纳为三大主要板块——办公主楼、变电站以及需要大跨空间的营业厅和调度中心，并针对用地界面的不同属性综合求解，以求将这三大功能板块精准落位。变电站和调度中心屏蔽了西北两侧城市道路的嘈杂，与塔楼围合出的南向内院与东侧的景观带声气相通。由城市道路经由南侧小路进入内院，绿化景观、入口水景、建筑门厅这一渐进的环境变化，将空间品质逐步提升至一个高潮。这看似平平无奇的归纳演绎过程，对外创造出了别致的建筑形体组合，对内满足了对景观资源的利用，实现了对内外环境的把握与控制。

在建筑形象设计上，务实、理性、亲民的建筑观贯彻始终。塔楼的立面设计更为注重城市视角下的建筑形象，方正的建筑形体、颀长挺拔的比例、严谨细腻的表面肌理，无不表达着理性、克制、自省的态度，业主所期待的行业形象得以重塑。而除此之外，面向城市道路的调度中心——电力行业的核心功能，在立面设计中特别加以强调彩釉玻璃上的数字化点阵，呈现出自由、交织的网格肌理，凝练而抽象地表达出“网”与“光电”的意象，以优雅、含蓄的方式展现现代电力行业建筑的创新形象。

从宁波城乡建委大厦、宁波城投大厦到宁波电力大楼，项秉仁在面对这些几乎是同时期设计建造的，性质和规模又几近相同的项目时，总能从分析项目所处的城市环境、基地条件、未来使用者的专业特点和使用要求着手，并由此寻找具有项目个性的设计出发点，然后展开步步深入的设计过程，直至项目的实施细部和现场执行。宁波电力大楼最为难能可贵之处在于，它彻底摆脱了民众对电力行业建筑挥金如土的看法，呈现出沉静而典雅的气质，含蓄而谦逊的态度。极为克制的创作令人耳目一新的同时也体现出行业进步发展的特质。建筑不炫富不浮夸不骄矜，融合于宁波城市建设的大背景之中，为周边空间的品质树立了标准。

1 营业厅入口
2 紧急疏散口
3 办公主入口
4 消控室独立出口
5 营业厅
6 休息等候候区
7 服务台
8 售餐窗口
9 备餐
10 领餐盘处
11 食堂
12 包间
13 驾驶员休息及更衣室
14 门厅
15 电梯厅
16 物业管理用房
17 消防控制与监控室
18 电镀层
19 主变坑
20 警卫室
21 安全工具间

一层平面图
First floor plan

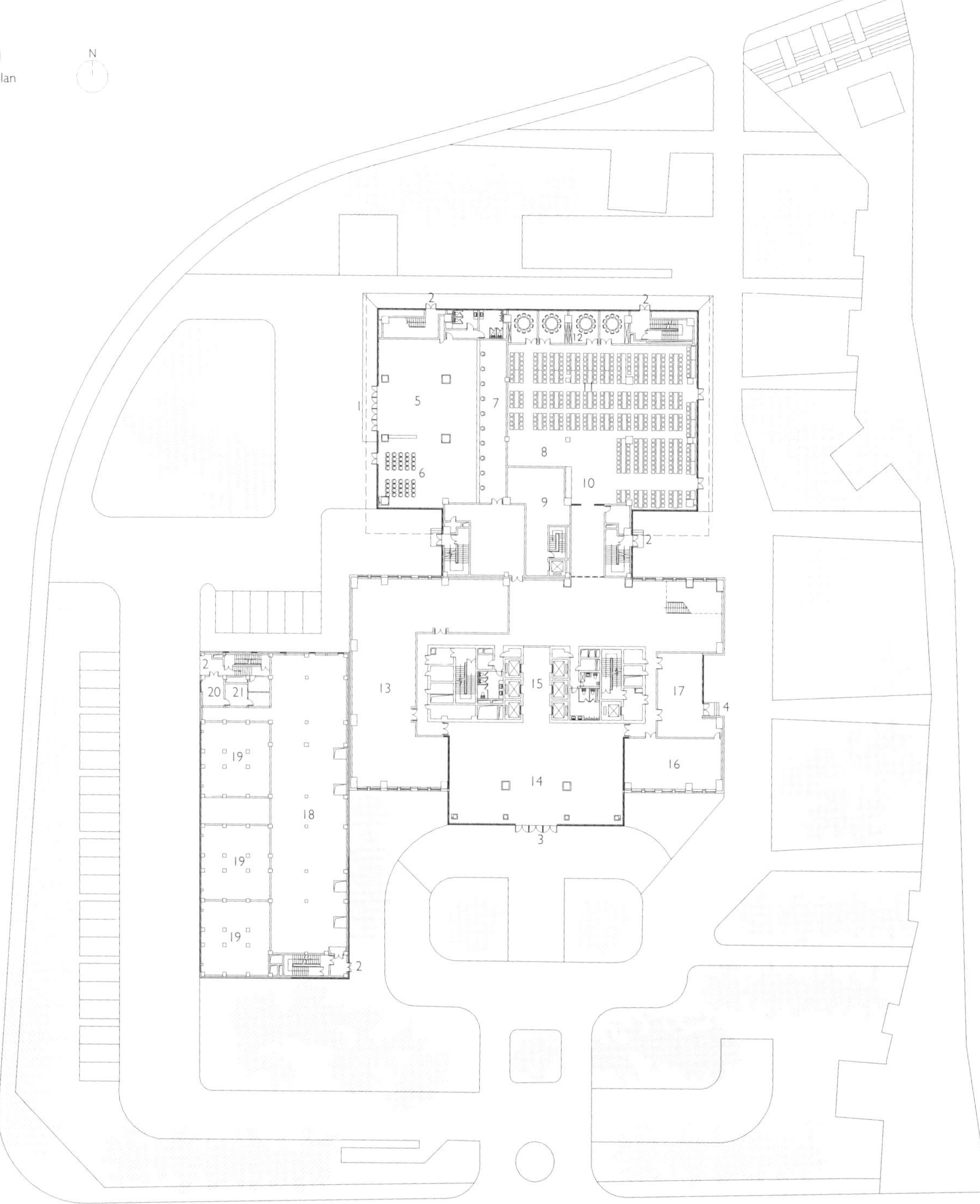

Project

# Ningbo Electric Power Authority Building

建筑实景 Building exterior view

The success of a series of administrative office buildings in the eastern part of Ningbo City won Xiang Bingren widespread praise in the field of public building design in Ningbo. Mr. Xiang's advocacy of democracy, transparency, openness and popularity in architectural design has combined with his rational and elegant architectural image to successfully subvert the stereotypes of traditional administrative office buildings. At the time, the Ningbo Electric Power Bureau had begun plans to build a new office and technology tower in Haishu District and they were eager to eliminate the public's negative perception of the firm as "a power monopolist", a goal that matched Mr. Xiang's design philosophy perfectly.

The site is an irregular rectangle with city roads on both sides of the west and north, a piece of commercial office land on the south end, and a 30-meter wide landscape buffer zone between the site and the river on the east. The characteristics of each interface were very clear. After interpreting the design task, Xiang Bingren divided the clients' functional demands into three categories: the main office building, the substation, and the business hall and dispatching center which required large spans of space. To insure that the programs were placed correctly, a comprehensive understanding of the different attributes of the above-mentioned interfaces was essential. The substation and dispatch center effectively shielded the noise from nearby roads on the west and north sides, creating an inner courtyard with a tower while also connecting it to the landscape on the east. Visitors are guided through gradual environmental changes from urban streets to inner courtyards via the south side road and green landscapes, where they are then treated to water features and hallways which culminate in a spatial climax. This seemingly mediocre induction, deductive process, creates a unique combination of architectural forms while fully utilizing the landscape resources,

建筑细部 Building details

achieving the control and shape of internal and external environment.

In the design of building facades, Xiang Bingren upholds a pragmatic, rational, people-centered concept. The design of the tower's facade focuses on its overall impact on the city. The cubic shape, the slender proportion, the precise and delicate surface texture all reflect the attitude of rationality, restraint and self-examination, fulfilling the client's intention. In addition, the facade design, which faces the main street, specifically emphasizes the core functions of the power industry as a dispatch center. The glazed glass facade of the center is covered with a layer of digital lattice, creating a free, intertwined grid texture. Through a concise yet abstract expression of the notions of "net" and "photoelectric", the design subtly and elegantly reveals an innovative image of modern power industry.

From the Ningbo Housing & Urban-Rural Committee Building, to the Ningbo City Investment Building and then the Ningbo Electric Power Authority Building, we found that when dealing with these office projects of almost the same size, Xiang Bingren always starts from analyzing the project's urban context, site conditions, requirements for future users, and unique characteristics. Only then does he proceed to carry out an in-depth design process step by step until the project's implementation. The biggest achievement of the Ningbo Power Building is that it powerfully shapes the public's image of its influence, showing a calm and elegant temperament along with an implicit and humble attitude. Moderate design is refreshing, but also reflects the progress of the industry: avoiding flashy. exaggerated or arrogant displays. The building is well integrated within the larger, urban context of Ningbo's development process and sets the standard for surrounding spaces.

## 评论
## Review

相邻时代的审美观虽然不同，却总有很多相似之处。向后看的观念传承与向前看的外部压力复杂交织，累积久了，只看头尾，即知巨变。对于任何一位跨越若干时代的建筑师，摆脱大众惯有的对于环境演变的不自知并不难；但能够捕捉身处当下的细微变化，拿捏向前与向后的微妙平衡，则需要天赋与经验，尤为难得。

有一天晚上项老师打电话来，说主楼立面正中央不是窗扇而是一片窄墙，言语中透露出担忧。那一稿设计采用了不规则排布的窗扇设计，三五种不同规格的窗扇在立面的水平与垂直两个方向上错动，形成匀质的网格。如果当时项老师没有提醒我，也许我并不会注意到他所说的问题。但延续当时的手法，是难以让窗扇始终居中的。设计立面的年轻同事觉得很委屈，认为方案并无不妥。事后看来，他们的分歧在于，两者都被向前的动力推动着，但项老师更清楚观念的传承对设计的影响。

史晨鸣

上海林同炎李国豪土建工程咨询有限公司 ANT 建筑创作与研发中心主任、主持建筑师

Though different, aesthetics of adjacent eras share many similarities. They move back and forth due to common external influences and the intrinsic need to inherent. Nearly unnoticeable mutations accumulate over time. Only by examining the beginning and the end do we realize how enormous the changes actually are. For any architect who has experienced the vicissitude of several eras, it is not difficult to escape the state of unconsciousness, like the mass public, towards an environmental evolution. However, to capture the subtle changes in the present, to grasp the delicate balance between forward and backward, requires talent and experience, which are especially rare.

Prof. Xiang called me one night, expressing his concern over how the center of the front facade of the main building was a narrow wall instead of a window. The design draft at the time features irregularly arranged windows: three to five different types of windows are staggered either horizontally or vertically, forming a homogeneous grid on the facade. If Prof. Xiang hadn't brought it up, chances are that I wouldn't have even noticed the problem and would have pursued the original design strategy, leading to uncentered windows. The young colleague who designed the facade felt frustrated, thinking that there was nothing wrong with the design. In hindsight, the divergence in opinions was caused by the designer's constant forward momentum in the absence of an awareness of the influence of the inheritance of ideas on the design. At this moment, however, Prof. Xiang proved the wiser.

Shi Chenming

Director/ Chief Architect,

ANT Architectural Creation R&D Center, Lin Tung-Yen & Li Guo-Hao Consultants Shanghai Co., Ltd.

西南向空间效果 South-west view

# China Merchants Group Shanghai Center



2012 / 2016

# 招商局上海中心

招商局上海中心地处后世博板块央企聚集区的腹地，项目源起于一块并不起眼的L形用地，建筑限高50米且要求低于毗邻地块，沿街展示面较窄。作为诞生于1872年的中国第一家民族工商企业，招商局在中国近现代历史上有着举足轻重的地位。塑造一栋富有个性且符合其百年传承的企业文脉，并能屹立于周遭林立的高层楼宇中的现代央企总部大楼是项目设计的目标，也是难题。

结合地块北侧的公共绿地资源，以及西南侧的内院资源，高层办公塔楼被设计成严整的L形，以顺应并最大化地利用基地条件。这一看似简单务实、毫无悬念的处理方式，恰恰蕴含了项秉仁对城市设计的理解尊重以及对于现代办公空间高效性的推崇。我们可以发现，这种理性思考在他之前的一系列办公作品中也是一脉相承的。

办公楼的交通核心布置于L形平面的阴角处，形成东、西两个较完整的通用办公区域，可分可合，为将来各单位的灵活划分提供可能。而在理性高效的建筑平面逻辑之下，贯穿着一个丰富多元的室内公共空间系统。建筑内设置的三个空中花园，以及一系列露台及屋顶花园将办公和绿色融合到一起，创造了更人性、更舒适的空间体验。通高的垂直绿化贯穿2层到6层，增加了各层之间的联系，形成了人与环境的互动。而10层与11层设置的空中花园和悬浮酒廊，为企业高层办公提供了静谧的休息和交流空间，高效的办公空间与灵动的空中园林相映成趣。

整个建筑以一种典雅厚重之势，温和淡薄之态端视周围。建筑学经典的三段式立面构图得以含蓄地表达，建筑呈现出“刚”与“柔”两种性格。正交格网的石材立面设计，从古典建筑中找到灵感，并提炼出理性和秩序，呈现出“实在”；转角玻璃界面的设计则采取最为轻盈的悬索张拉幕墙的方式，尽可能地非物质化，呈现出“轻无”，并使空中花园体系得以在立面上隐约地表达。两种立面表达戏剧化地铺陈并置，以高度统一的造型元素，内外相应的空间策略，一气呵成地塑造了这座理智与热情并置的城市新地标，诠释着招商局“百年传承，历久弥新”的企业文化。

而对建筑细部的高度关注，也体现出一种务实、耐性和坚守的匠人情怀。石材的划分以及到转角的拼接都有细心的考虑，石材与玻璃之间以毫不含糊的精致边框收头交接。对建筑学的追求深入到每一处胶缝和每一种交迭的细部之中，这也使得这座建筑可以成为不骄不躁、能与时代并肩而行的典雅之作。

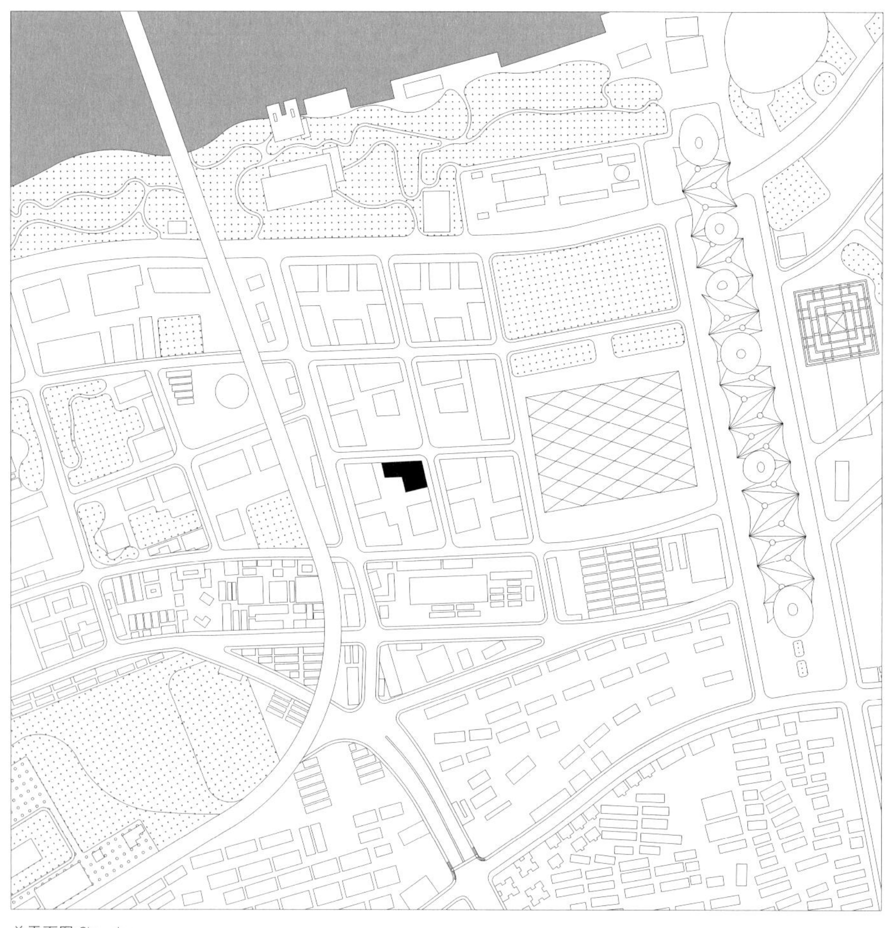


总平面图 Site plan

The China Merchants Group Shanghai Center is located in the heart of the central enterprise gathering zone in the post-expo area. The project area is shaped like an "L" resulting in a short street display. Its height limitation is 50m lower than neighboring buildings. Founded in 1872, the China Merchants Bureau was the first national industrial and commercial enterprise of China and has long held this pivotal position. One challenge for the new center design was how to integrate the headquarters for this unique modern central enterprise with the surrounding high-rise buildings.

The high-rise office building is designed to be L shape to follow the site shape as well as to maximize the utilization of the northern public green and southwestern inner court areas. The design appears simple and suspense-free, but we can find the urban design is fully understood and respected by Professor Xiang Bingren as he praises the high efficiency of the modern workplace. This deliberate, measured thinking characterizes all his previous workplace achievements.

The office building includes west and east wings as general office space which can be divided or united freely according to the needs of different companies who are going to use the building in the future.

Rational and high efficiency characterize the whole interior public space.

There are three hanging gardens and a series of terraces and roof gardens in the building which integrate the office space with landscaping for a more humanized and comfortable experience. The vertical green wall connects the floors between 2F and 6F and makes people interact

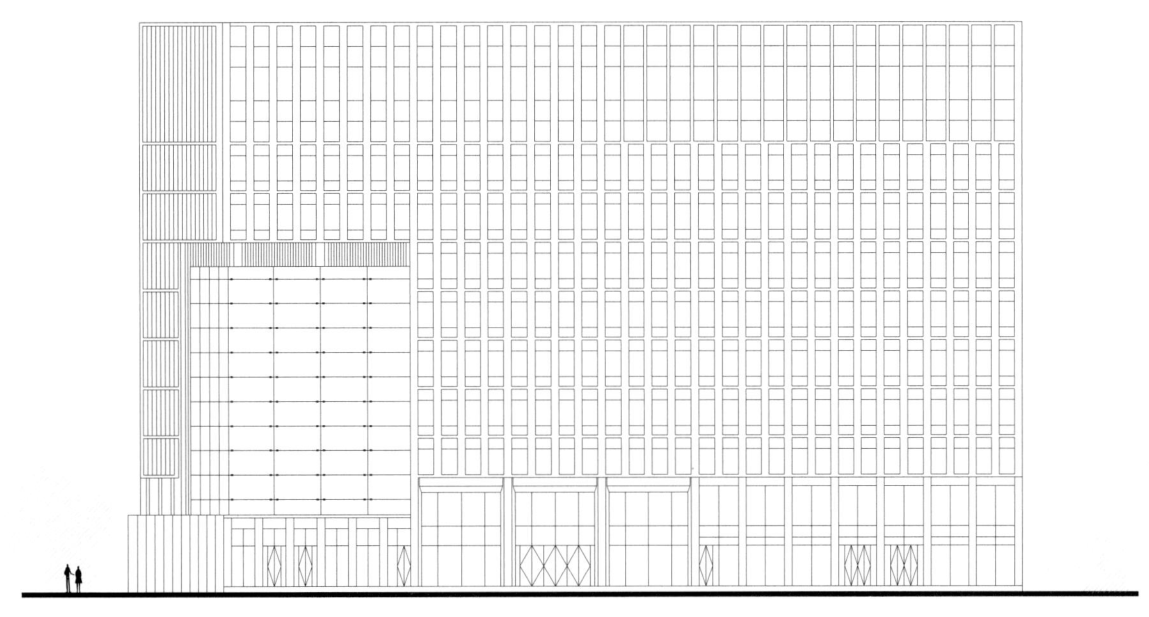

北立面图 North elevation

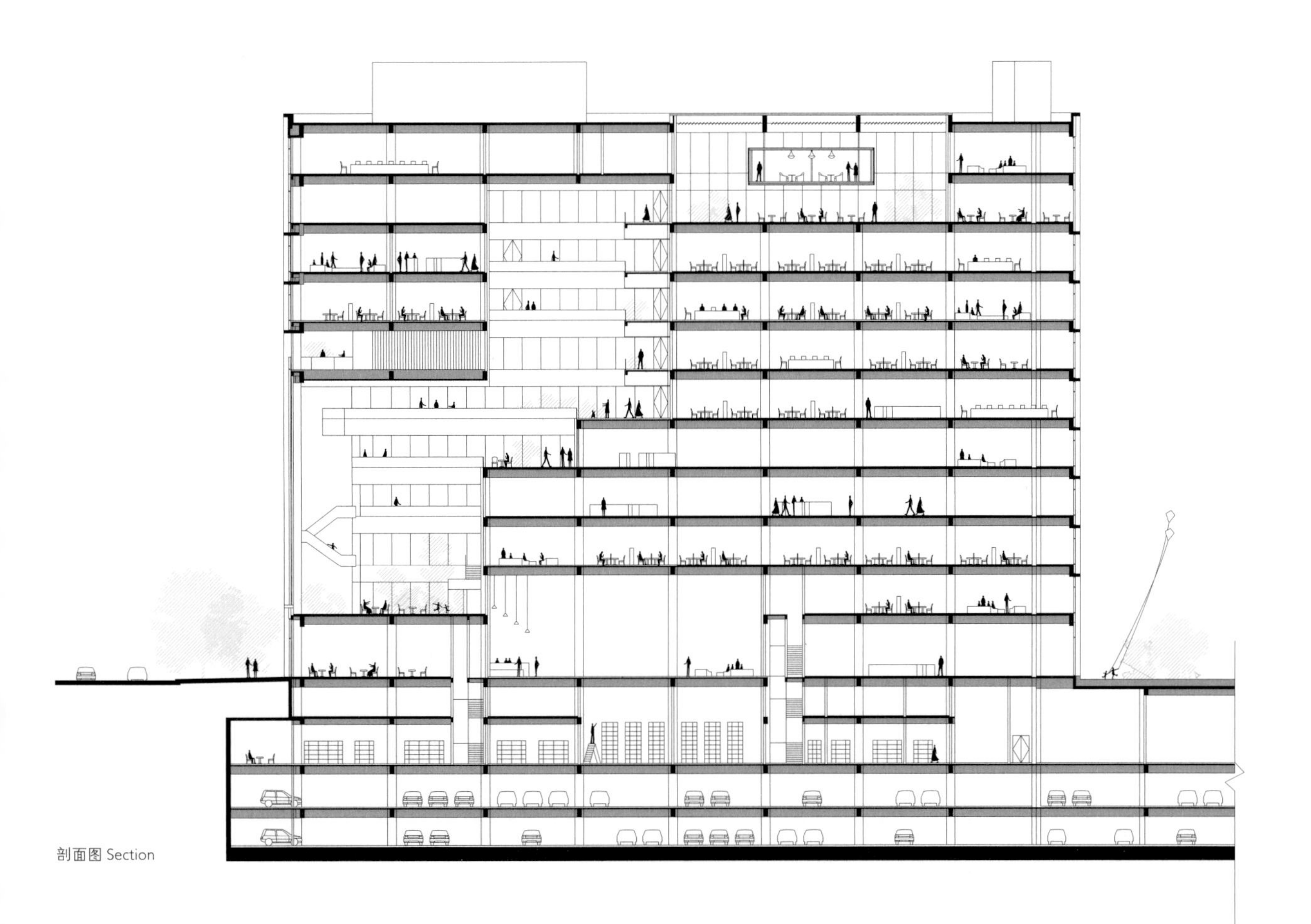

剖面图 Section

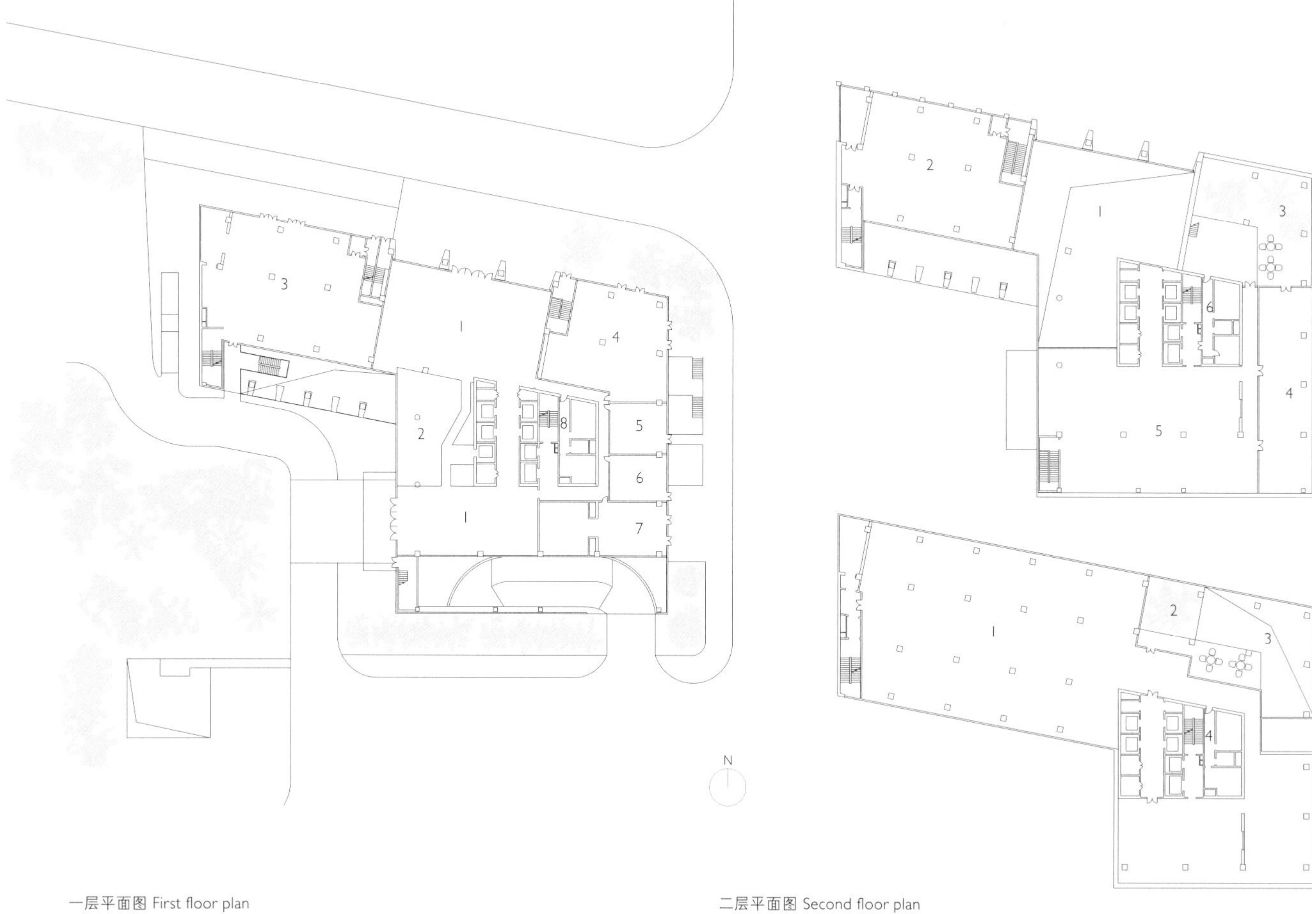


一层平面图 First floor plan
1 门厅大堂 2 景观水池 3 商业（招商银行） 4 商业（咖啡厅）
5 商业（票务） 6 商业（便利店） 7 商业（蛋糕烘培店） 8 卫生间

二层平面图 Second floor plan
1 大堂上空 2 办公（招商银行） 3 空中花园
4 室外平台 5 办公（物业办公） 6 卫生间

五层平面图 Fifth floor plan
1 办公 2 空中花园
3 花园上空 4 卫生间

with nature closely while inside the building. Occupants can rest and converse in the sky garden and floating lounge on the tenth and eleventh floors. This interaction between high efficiency office space and flexible sky garden has been well received by visitors and regular occupants.

The building takes an elegant while steady look and expresses mild and indifferent attitude toward its inferiors.

There is a modest expression of classical architectural ternary form on the facade endowing the building with two characteristics: hardness and softness.

The orthorhombic lithoid facade expresses "real" inspiring and extracting from classical buildings; corner glass is lightest suspension cable glass curtain wall and expresses "soft" making the hanging garden implied.

The two dramatic juxtaposition of the two facades, with highly uniformed elements and internal and external corresponding spatial strategies, forms a coherent whole new urban landmark full of reason and passion while interpreting the CMPD culture "never fading charm and legacy lasting over century".

The highest attention is given to design details in the spirit of craftsmanship: pragmatic, patient and insistent. The division of stone and the split joint at the corner are given particular attention. This intense focus also creates an exquisite frame between stone and glass.

The serious architectural attitude can be seen from every glue line and every detail crossover. Such an attitude makes this building a classic and yet congruent with modern sensibilities and avoiding arrogance and impatience.

百年传承历久弥新央企的现代建筑形象 A contemporary building image

夜景效果 Night view

空中花园的幕墙设计带来明亮的光线 Bright light of sky garden

## 评论
## Review

建筑从其诞生的一天便是时间的朋友。无论是4000多年前的古埃及金字塔，还是建造了一个多世纪至今仍未完工的圣家堂，无不在时光的洗礼下焕发出经久的光华。然而在疾风骤雨般的当代中国建筑实践中，做时间的朋友似乎成了一种奢望。
所幸，招商局上海中心是一个成就与时间共舞的建筑的契机。从2012年夏到2016年秋，DDB经历了近十轮的方案研讨、几十次设计的对接会议、三版幕墙样板、十几次施工现场对接、多轮设计调整：一边是各工种设计方案的不断优化、推翻和演进，一边是业主对设计的尊重和对细节实施的坚持和认真。四年来共同的努力、汗水、辛酸和遗憾汇在一起，五味杂陈，无不值得我们回味和深思。虽然有种种遗憾和不足，但是这样一份答卷使我们无愧于心。
建筑设计实践是一门高度复杂和综合的艺术。好的项目的实现需要的不仅是设计师的积累、磨练、传承和创新，还要求我们与开发者、合作者们共同建立起与时间同行的理念。

马庆禕

上海秉仁建筑师事务所首席合伙人、总建筑师

Architecture has always been an intimate cultural concern since it was first conceived. The pyramids, built 4000 years ago, and Sagrada Familia which has been under construction for more than 100 years are crowned with timeless beauty by the erosion of time. However, it seems to be quite luxurious trying to associate architectural practices with time in China nowadays due to the fanatic booming of construction.
Luckily, the project of the Shanghai headquarter of the China Merchants Group is a precious opportunity where a piece of architecture that dances with time was designed and erected. From the summer of 2012 to the autumn of 2016, we raced with time for 4 years. We went through nearly 10 different proposals, dozens of meetings with the clients, 3 versions of curtain wall design, dozens of negotiations with the constructing team and countless adjustments to the design. We tried our best to optimize, to evolve even to overturn the design in order to pursue perfection. At the same time, the client showed great respect toward the design process, and showed sincerity, respect and persistence to the implementation of each and every one of the details. We deserve to bask in the feeling of a job well done after four-year hardworking. Although no work of art or design is perfect, we stand proudly by the results of our dedication and vision.
Architectural design practice is highly complex and integrated art. A good project more than the designer's experience, inheritance and creation alone, it also demands the same idea between designer, developer and collaborators.

Ma Qingyi

Chief Partner/ Executive Chief Architect, DDB Architects Shanghai

# 5

## 大型文化建筑：回归建筑学本质

文化建筑作为人类文明的标志，是一个国家或民族的历史和现代人文的物质结晶。自20世纪末以来，文化建筑又因其具有代表城市文化竞争力的特质，在国际舞台上呈现出不断增多的态势。正是在这样的背景下，项秉仁相继为合肥、西安、宁波、金华等城市设计了重要的地标性城市“文化名片”。

以合肥大剧院（2003）、西安大唐不夜城贞观文化中心（2005）、宁波文化广场（2008）三个项目为典型代表，可以看出项秉仁对于这一类型建筑的创作已经完全跳出了各种主义与学理的羁绊。他从城市环境出发，由外而内地针对具体问题寻求最佳设计答案，以服务业主、满足使用者预期为指导原则，以平衡功能、实现高完成度的作品作为最终目标。在某种意义上，项秉仁的建筑创作已经获得了新的自主。这种自主表现在设计方法与工具的不拘一格上，也表现在形式、功能与意义的张力上。

## Large-scale Cultural Buildings: Return to the Essence of Architecture

Cultural architecture has always been a symbol of human civilization, and since the end of the 20th century, it also takes up the task of “city branding” and has become an increasing trend on international stage. Chinese cities are not far behind, and Xiang Bingren has become the brain behind several cultural landmarks for cities like Hefei, Xi'an, Ningbo, and Jinhua.

Hefei Grand Theatre (2003), Great Tang Everbright Town Zhenguan Cultural Center, Xi'an (2005) and Ningbo Cultural Plaza (2008) reveal that Xiang's approach towards this type of building has completely liberated from various doctrines and theories. Starting from the urban environment, he seeks the best design solutions to specific problems from the outside to the inside. He practices on the principle of serving the clients and meeting end users' expectations, and the ultimate goal is to balance various functions and achieve high degree of realization. In a sense, Xiang's architectural creation has gained a new autonomy, which can be seen in his free choice of design methods and tools, as well as in the tension between form, function and implication in his buildings.

剧院鸟瞰 Bird's-eyes view

2003/2009

# 合肥大剧院

2003年安徽省合肥市政务文化新区为拟建的合肥大剧院发起国际设计方案竞标，项秉仁和他的设计团队受邀并提出设计投标方案，最终他的设计被选定为中标方案并予以实施。

合肥大剧院的建筑形象带有强烈的浪漫主义色彩，但从总体规划到建筑单体设计的整个过程，又蕴含着许多必然的内在理性和生成逻辑。在城市区位上，位于政务新区的规划科技轴线两侧的大剧院和艺术馆交相呼应，顺应已有的对称城市空间形态，且与前期设计的"行政中心"在平面构图上形成稳定的三角关系。结合南侧的天鹅湖，这四个城市环境控制性要素塑造了稳定的内聚城市中心，柔化的建筑界面与南侧湖景资源融为一体。这样的整合使大剧院不仅仅只是市民享受艺术的场所，其本身也成为城市的一件艺术品。

建筑造型的灵感源于天鹅湖的水纹，由"水纹"建筑母题衍生出的圆润柔性轮廓线，赋予建筑形态一种飘逸且沉静的效果，和天鹅湖畔静谧的自然环境相得益彰，宛如一颗熠熠生辉的"湖畔明珠"。值得一提的是，为了解决大剧院和艺术馆两者对称而又体量悬殊的棘手问题，它们各自采用了正负高斯曲线的形式以相互平衡。大剧院采用正高斯曲面，形象舒缓内敛，以消解巨大的尺度对比；而艺术馆采用负高斯曲面，呈现强烈的向外伸展动态，在视觉上达到扩张。再进一步，圆弧形的屋面构造根据不同的水平模数化做切片处理，统一匀质的肌理处理取得一定程度上的均衡对称的效果。不同的弧形切片层层交叠、跌宕错落、起伏有致，以丰富的形态将公众大厅及休息前厅向湖面展示，晶莹剔透的玻璃幕墙及银色钛金板屋面在湖面形成优美的倒影，创造出自然、和谐、优美的意境。

换个角度来看，合肥大剧院如此浪漫随性的建筑外形还有其内在生成因素。曲面屋顶所覆盖的空间包括共享大厅、歌剧厅、音乐厅、多功能厅以及其他附属设施。各个不同空间的需求与大剧院的椭圆轮廓正好契合，恰如其分地满足了三个演艺厅的空间与舞台高度的需求。屋顶水平切片的处理提供了室内空间的可塑性，弧形切片之间形成的错落还很好地解决了侧窗采光通风的问题，整个公共大厅空间开阔而富有变化，极富流动性。

建筑主入口 Main entrance

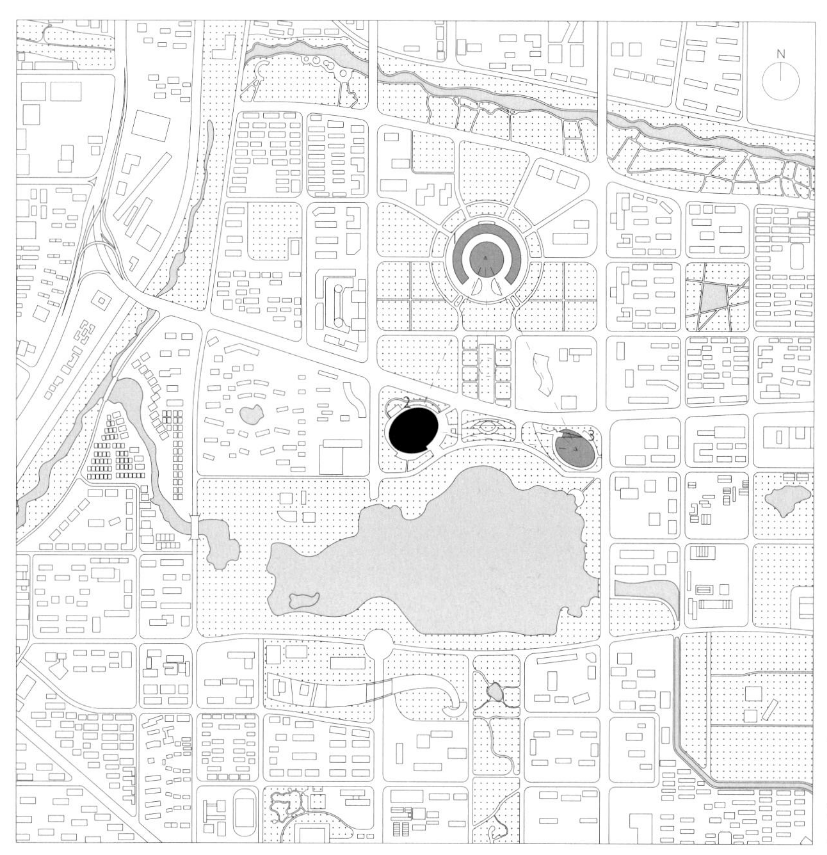


总平面图 Site plan

1 合肥市行政中心
2 合肥大剧院
3 美术馆（未建）

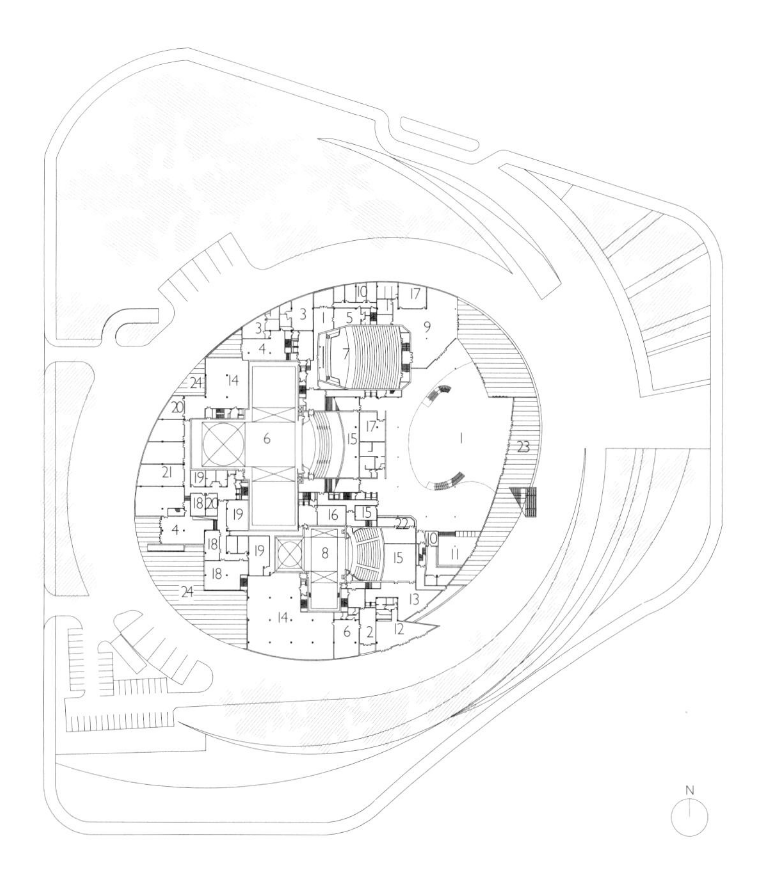


一层平面图 First floor

1 入口门厅
2 贵宾入口
3 演员服装间
4 员工 / 演员进厅
5 贵宾休息室
6 主剧场
7 音乐厅
8 戏曲艺术中心
9 管理室
10 办公
11 售票厅
12 餐厅
13 餐厅入口大堂
14 布景组装 / 存放
15 设备房
16 演员休息室
17 吸烟室
18 服装室
19 休息室
20 库房
21 更衣 / 化妆
22 寄存处
23 室外平台
24 卸货平台

室内大厅与旋转楼梯 Main lobby

In 2003, Hefei New Municipal and Cultural District in Anhui Province initiated an international RFP for the future Hefei Grand Theatre. Xiang Bingren accepted the invitation and lead his design team to work out a proposal. Finally, his design was selected and implemented.

The architectural image of Hefei Grand Theatre possesses a strong romantic essence while the overall planning and architectural design contain many intrinsic yet necessary features. The Grand Theatre and the Museum of Art are located on either side of the science and technology planning axis in the new district. They echo the existing symmetrical urban space while forming a stable triangular relationship with the existing "administrative center". Combined with the Swan Lake on the south side, these four urban elements control the local environment while also shaping a stable and cohesive urban center. They additionally soften the architectural interface that blends with the Lakeview resources on the south. This integration raises the Grand Theatre above its basic function as a venue for the public to enjoy art. It has itself become a piece of art in the city.

The building's mellow and soft silhouette carries the theme "Wa-

歌剧厅实景 Opera house interior

termarking", inspired by Swan Lake. The thematic choice provides for the building an elegant and tranquil temperament that is in harmony with the peaceful natural setting of Swan Lake, like a shining "lakeside Pearl". One challenge encountered by the design team was how to approach the drastically different volumes between the Grand Theatre and the gallery, which stand in symmetrical positions. The design adopts the form of positive and negative Gaussian curves to balance the difference. The Grand Theatre adopts a positive Gaussian surface with soothing and introverted images to break down the massive scale. The Museum of Art adopts a negative Gaussian surface, exhibiting a strong outward visual expansion. Further, the arcuate roof structure is sliced according to different horizontal modular numbers while the uniformity of the surface texture achieves a certain symmetry. Different arc-shaped slices overlap each other generating an up and down visual momentum, showing the public lobby and lounge lobby to the lake through its rich form. A crystal clear glass curtain wall and silver titanium roof form a beautiful reflection on the lake to create a natural harmonious scene.

From another point of view, the Hefei Grand Theatre's romantic yet

交响乐厅内景 Symphony hall interior

casual style is derived from some intrinsic factors. The roof covered spaces include a shared hall, a 1500-seat opera house, a 1000-seat concert hall, a 500-function multi-functional hall, and other ancillary facilities. Spaces for various purposes perfectly fit into the elliptical outline while meeting the space and stage height requirements of the three auditoriums. The processing of the roof's horizontal sections provides the plasticity of interior space. The scattered formation between sectionalized arcs also solves the need for illumination and ventilation through the side window. The public hall contains a wide and varied space that is very fluid.

High-platform was made up of double-skin facades. The outer was faux stone metal grille, and the inner was glass curtain wall and concrete wall. This merged modern materials technology with modern aesthetics to show the spirit of the region. The upper area for high-platform designed to watch shows also adopted double-skin facades as the elevation. The outer used faux wood aluminum alloy to coordinate with the Tang Dynasty wooden architecture. Bays and windows embodied traditional Tang style.

从天鹅湖望大剧院 Lake view of the theatre

## 评论
## Review

当时正是大量西方建筑师涌进中国垄断各种重要标志性建筑设计权的前夜，设计生态和大众审美都在逐渐发生演变，但相对来说还是给本土设计师留下了一些公平竞争的机会。这个设计能够赢得竞赛也是一种改良的本土设计策略的成功，其目标非常诚实地反映了设计者在专业创新和公众意识之间谋求平衡的意图，整个操作过程则在希望接纳新的非线性形式范式与尚在发展中的数字化设计技术之间纠缠，最终结果也非常诚实地表现出多元化审美需求与传统的美学规则之间的博弈，应该算是在数字化设计普及所带来的新形式井喷之前的一次卓有成效并充满意味的尝试。从这个意义上来说，其所追求的建筑愿景已经脱离了传统的“美”与“丑”的二元判断，取而代之的是更为模糊、更加深层次的审美情感体验。

董屹
同济大学建筑城规学院副教授、硕士生导师
DC 国际建筑设计事务所合伙建筑师

合肥大剧院整体布局形态的空间逻辑体现了建筑师在城市设计维度的深层次思考。大剧院位于新区科技轴的两侧，与中轴线对面的综合艺术馆共同构成合肥文化艺术中心最重要的两个单体。建筑师最终选择了两个成角度的椭圆作为建筑的主体轮廓线，使其在轴线上的交汇点落在城市主轴上，同时将主轴的端部作为建筑对外的主要入口空间，这样既解决了两个建筑在城市主轴上的对位关系，同时其曲线的形态也与天鹅湖形成有机的整体。在一个椭圆球体的基础上，两个建筑在立面上的曲线被处理成正反两个方向，这样使得两个不同体量的建筑在整体上取得了统一平衡的关系。从城市空间形态视角，两个不同功能属性的建筑所形成的整体形态，共同衬托了位于中轴线远处的市政府大楼，同时自身也成为符合天鹅湖自然形态的背景建筑。

韩冰
天华集团资深合伙人
上海天华执行总建筑师
建筑学博士

It was a time right before numerous Western architects rushed into China and monopolized the design for many of the country's important landmarks. Though the ecology of the design industry and the public's taste were gradually evolving and constantly challenged, a few local designers were still fairly active in these design competitions. The fact that Mr. Xiang's design won the competition, against such a backdrop, suggests the success of modified local design. If we look closer, Mr. Xiang's design honestly reflects the designer's intention to strike a balance between professional innovation and the collective consciousness. Moreover, the designing process expresses the desire to embrace a new, nonlinear formal paradigm along with developing digital design technologies. The final result honestly reveals the competitive relationship between the needs for a diversified aesthetic and the formal rules favored by traditions at that time. At the time, it was a meaningful and successful example of design prior to the disruptions caused by the introduction of digital design technologies. In this sense, the architectural vision pursued by Mr. Xiang had broken away from the traditional sole criterion of "beauty vs. ugliness" and instead adopted an approach that has been described as providing a vague, deeper aesthetic and emotional experience.

Dong Yi
Associate Professor/ Master Supervisor, CAUP, Tongji University
Partnership Architect, DC ALLIANCE

The spatial logic of the Hefei Grand Theatre reflects the architect's urban-minded considerations. The Grand Theatre is located right on the Science and Technology Axis of the New District along with the Comprehensive Art Museum, across the axis. They are together, the two most important elements of the Hefei Cultural and Art Center. The architect chose two angled ellipses as the layout of the theater, whose axial intersection falls on the city's axis. The main entrance of the building is placed at the end of this main axis. This strategy works out the alignment between buildings and the axis while the curved form of the theater also forms an organic whole with Swan Lake. Based on the ellipsoidal sphere, the architects created curves of competing directions for the facades of the two buildings, achieving a visual equilibrium between the two buildings of different volumes. From the perspective of its urban environment, the two buildings function together to help frame the Municipal Building located at the far end of the axis. The theater itself has become a suitable backdrop for the scenic Swan Lake.

Han Bing
Senior Partner, Tianhua Group
Executive Chief Architect, Shanghai Tianhua
Doctor of Architecture

建筑细部 A close-up view of the center

# Great Tang Everbright Town Zhenguan Cultural Center, Xi'an



2005 / 2017

# 西安大唐不夜城 贞观文化广场

2005年项秉仁经过国际设计竞赛成功赢得了西安曲江新区大唐不夜城贞观文化广场的设计。曲江新区作为十三朝古都西安中心城区东南侧的新兴城区，当时已初步形成了以大雁塔为中心、具有浓郁唐代园林风格的旅游区。作为曲江开发的明珠，大唐不夜城将被打造成重现大唐盛世风采的文化商业步行街区，而其核心部分贞观文化广场项目，由大剧院、音乐厅、电影城和美术馆四组文化艺术型建筑组成，总建筑面积约为10.1万平方米。

由于项目所处大雁塔历史文化遗址的特殊环境，以及西安悠久的历史文脉，新建的项目一方面必须延续城市环境的整体风貌，体现古典传统建筑文化的精粹；另一方面又要满足当代演艺文化博览建筑的人流和功能需求，同时还要使新老建筑形态肌理和谐完美地融为一体。项秉仁将四个主体建筑以正对大雁塔的南北轴线为中轴对称布置，围合成唐代常见的“凹”字形空间布局，形成方整、内向、极具聚合力的现代文化交流空间。四个主体建筑的大空间及屋顶采用南北向布置，次要空间呈东西向布置，以顺应中国传统建筑的布局特点，并凸显四个文化建筑的特殊地位。大剧院和音乐厅主体屋顶采用重檐庑殿，美术馆和电影城主体屋顶采用重檐歇山，次要空间设计为大小不等的单檐歇山，既强调了四个主体建筑的壮观形象与气势，也体现了盛唐设计意象中的大小屋顶组合的建筑群落感，使新建的建筑群与整个大唐不夜城融为一体。

考虑到四个主体建筑从功能空间组织上都是观演展览功能的大空间结合附属设施的组合方式，建筑风格采用了唐代盛行的高台建筑的意象。附属设施纷杂的体量被统一在一个简洁、现代的高台下，而观演建筑的大空间则用原汁原味的唐风大屋顶建筑覆盖，放置在高台之上。屋顶从坡度处理、斗拱尺寸、出挑深度及细部设计方面都充分再现了唐朝屋顶的形式，体现了新建筑对原有地域空间特征的敬意。这种古今叠加的造型处理方式尊重传统优秀文化传承，摈弃了纯粹复古的建筑模式，并且拒绝含糊不清的语义，是把原汁原味的唐风建筑作为传统文化的精粹和信息载体，与具有现代审美理念的高台基座进行并置对话。

具有现代审美的高台由双层表皮构成，外层为仿石金属格栅，内层为玻璃幕墙与素混凝土墙体。这是利用现代的材料工艺，以当代的美学表现方式去体现传统地域风貌的精神特质；在高台之上的观演空间同样也采用了双层表皮的建筑立面形式，其中外层采用铝合金仿木的形式回应了唐代木构建筑的体系、开间和开窗方式，体现出传统唐风建筑的内涵。

# 西安大唐不夜城贞观文化广场



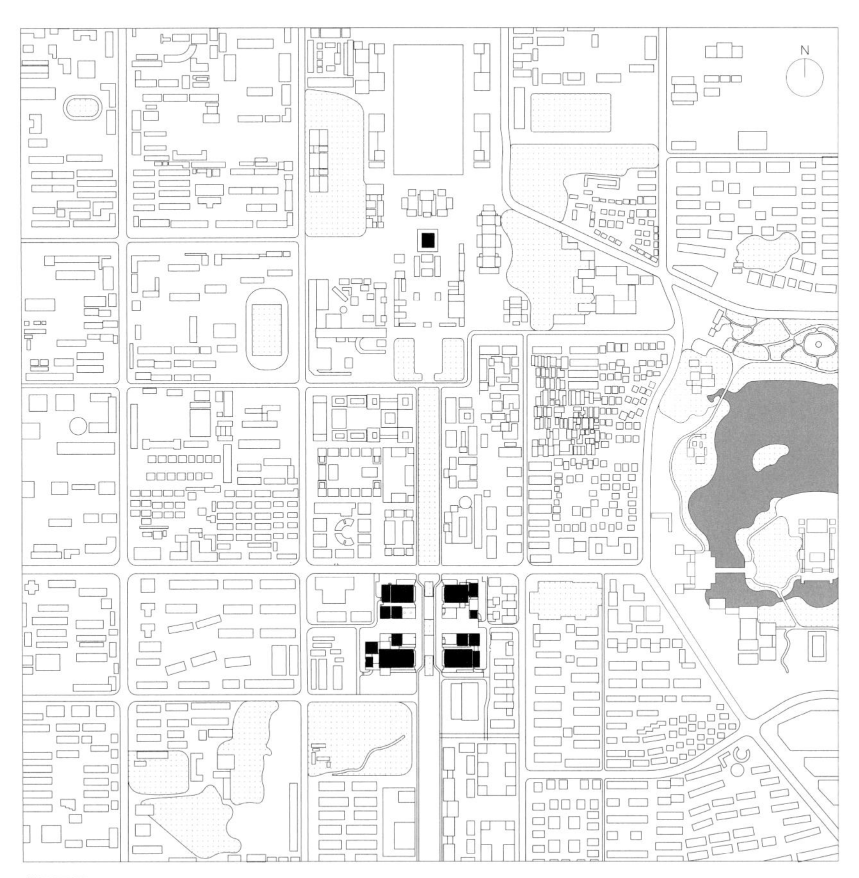


总平面图 Site plan

In 2005, Xiang Bingren won the international competition to design the Great Tang Everbright Town Zhenguan Cultural Center at the Qujiang New District of Xi'an. As an emerging urban area southeast of downtown Xi'an, the Qujiang New District, which was hosted the capital of thirteen ancient dynasties, has built up a Tang Dynasty-style tourist zone with the Dayan Pagoda as its center. As the center of Qujiang's development scheme, the Great Tang Everbright Town strives to recreate Tang-style pedestrian neighborhoods for culture and commerce. The Zhenguan Cultural Center, a key development project, includes four building clusters with approximately 101,000 square meters of total area: Opera House, Concert Hall, Film City and Art Museum.

The site's present location is immediately adjacent to the Dayan Pagoda and Xi'an's historical context as Chang'an, capital of Tang Dynasty. This placement means that the new project needs to both continue the overall style of the urban environment while embodying the essence of traditional architectural culture and also must fulfill the needs of contemporary concert halls, theaters and expo buildings, which have very strict circulation and functional requirements. Moreover, the new design interventions needed to perfectly and harmoniously integrate the new and old. Mr. Xiang arranged the four main buildings

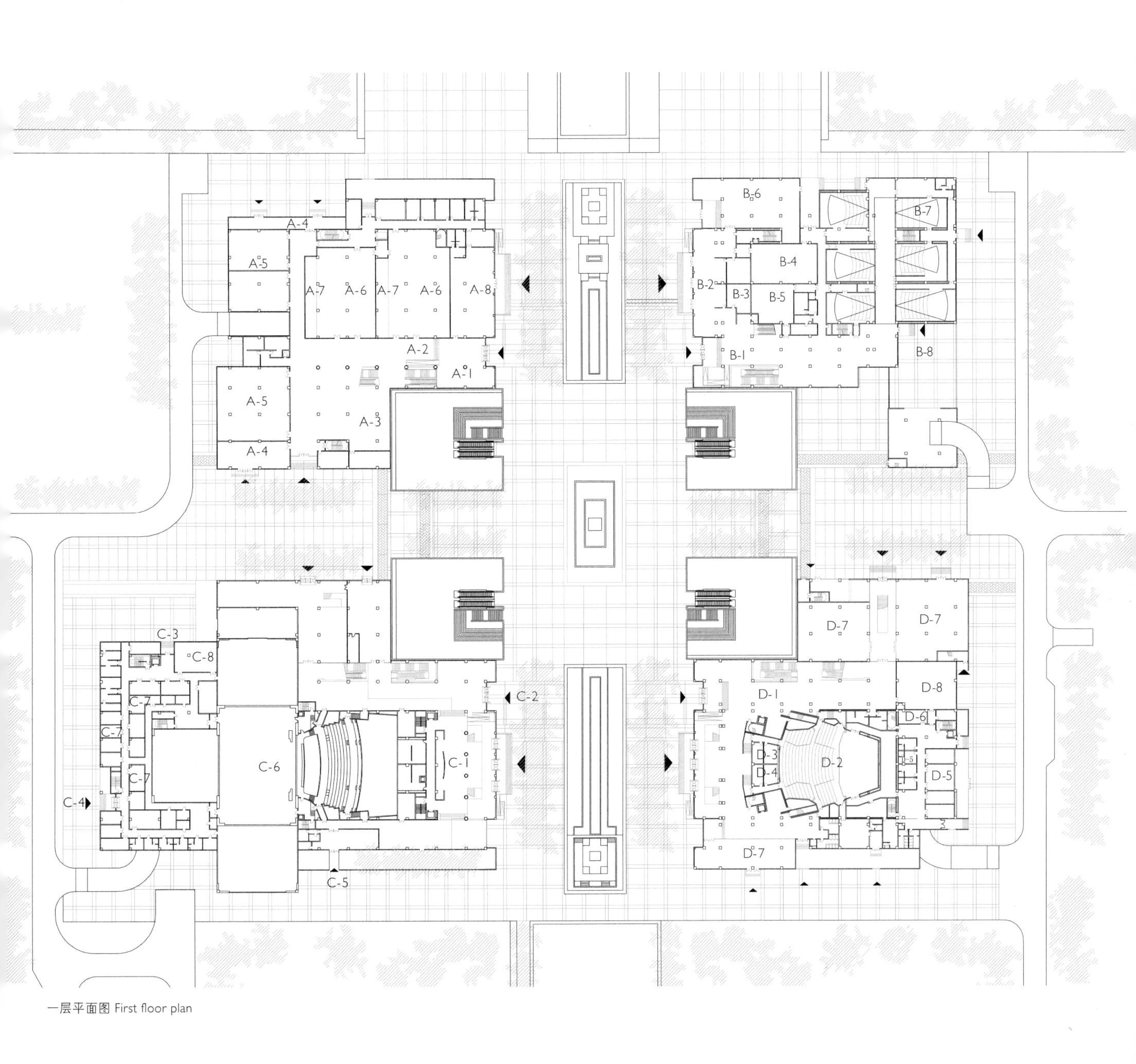


一层平面图 First floor plan

| A | 西安美术馆 | B | 太平洋电影城 | C | 陕西大剧院 | D | 西安音乐厅 |
|---|---|---|---|---|---|---|---|
| A-1 | 入口门厅 | B-1 | 入口大厅 | C-1 | 入口大厅 | D-1 | 入口大厅 |
| A-2 | 艺术书吧 | B-2 | 售票大厅 | C-2 | 疏散出入 | D-2 | 音乐大厅 |
| A-3 | 艺术商店 | B-3 | 票务中心 | C-3 | 布景入口 | D-3 | 音控光控 |
| A-4 | 艺术工坊 | B-4 | 多功能厅 | C-4 | 演员入口 | D-4 | 存衣设备 |
| A-5 | 文物鉴赏 | B-5 | 点播放映 | C-5 | 贵宾入口 | D-5 | 化妆休息 |
| A-6 | 休息区域 | B-6 | 卖品商店 | C-6 | 主舞台区 | D-6 | 贵宾休息 |
| A-7 | 休闲咖啡 | B-7 | 贵宾观影 | C-7 | 化妆休息 | D-7 | 音乐商店 |
| A-8 | 办公区域 | B-8 | 唐诗词园 | C-8 | 淋浴区域 | D-8 | 数字摄影 |

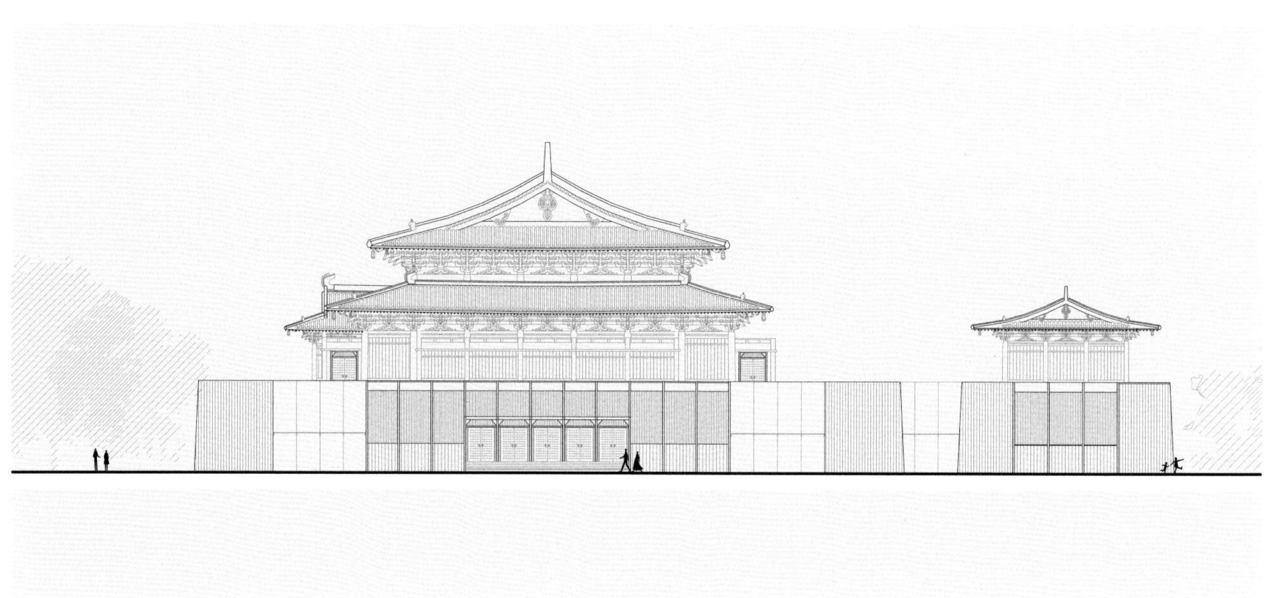

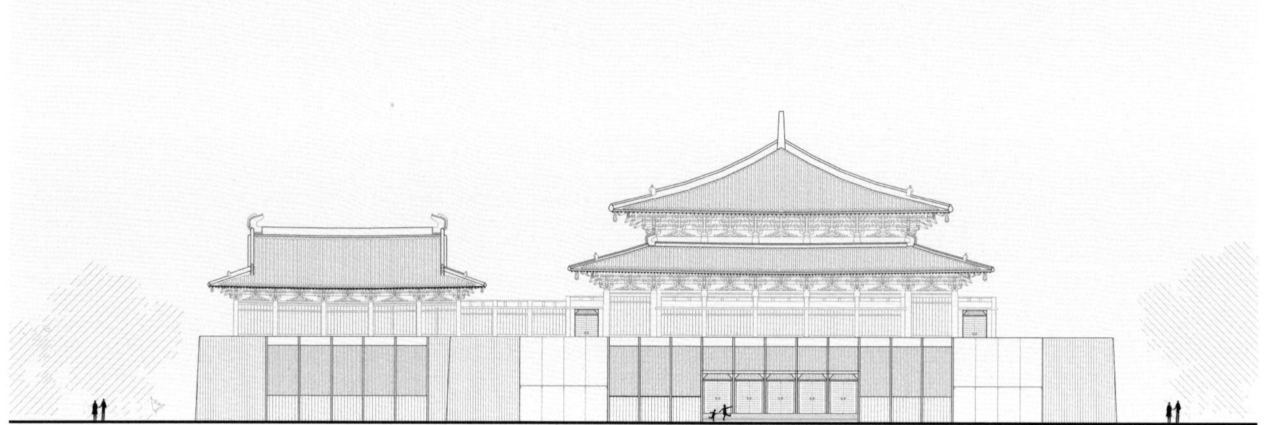

^ 太平洋电影城
∨ 西安音乐厅

立面图 Elevations

symmetrically on the north-south axis of the Dayan Pagoda, and formed a representative "凹"-shaped spatial layout in Tang Dynasty-style. The arrangement creating a neat, introverted, and highly convergent modern cultural communication space. According to the traditional Chinese spatial hierarchy, larger spaces and main buildings are arranged in a north-south direction while secondary spaces are east-west oriented, indicating a special status of the four cultural buildings. The main halls of the Opera House and the Concert Hall uses a traditional double eave hip roof. The main halls of the Art Museum and the Film City adopts the double eave gable and hip roof. As for secondary spaces, they all adopt the single eave gable and hip roof though of varying sizes. Such an arrangement serves to emphasize the spectacular image of the four main buildings while reflecting the representative Tang Dynasty preoccupation with aggregating buildings of many sizes. Also the special roof design helps this new construction to blend into the Great Tang Everbright Town.

Functionally speaking, the four main buildings are composed of larger performance spaces combined with ancillary facilities. In this way the design reinterprets the "hall on a high podium" building style which prevailed in the Tang Dynasty. It goes further and places various ancillary

^ 总体鸟瞰效果图 Bird's-eyes view

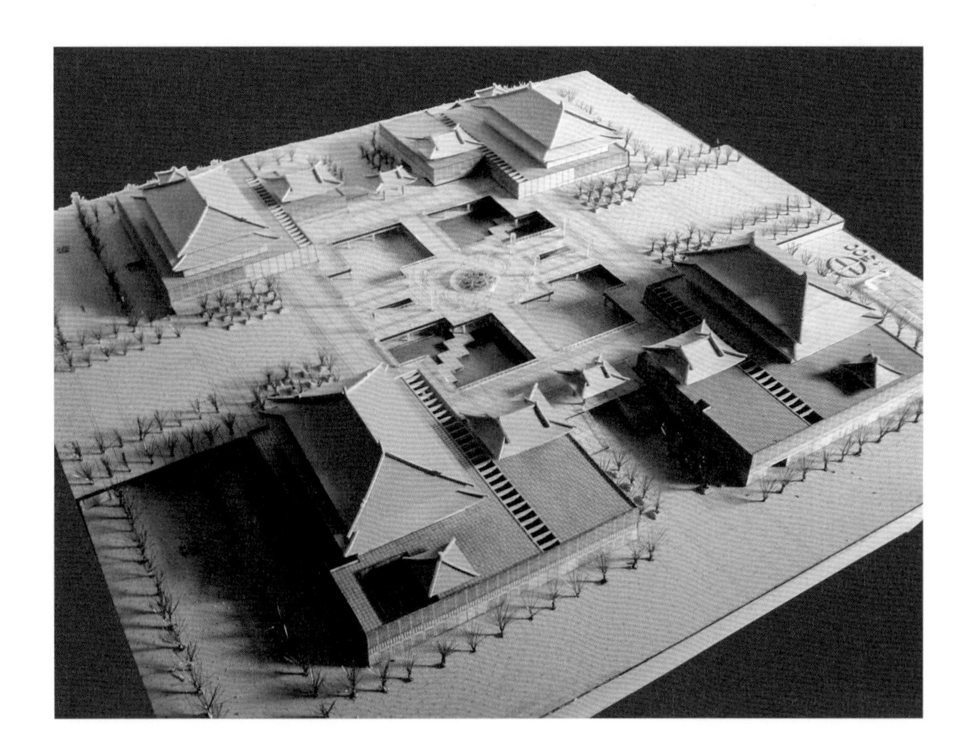

> 建筑模型 Model

现代基座与唐风屋顶的碰撞 Exterior view

facilities within a simple, modern podium, on which stands the main spaces under a large authentic Tang-style roof. The new design pays homage to the existing regional spatial features by strictly recreating the sloped roof, the bucket system, the eave's width and many detailed designs according to the Tang style. The strategy of juxtaposing the ancient and the modern both shows respect to the traditional cultural heritage while also avoiding literal reproduction and ambiguous expression. Instead, the strategy of utilizing the Tang-Style architecture as a carrier of traditional culture, creates a conversation with the podium that bears modern aesthetic features.

The modern podiums are covered in a double-layered facade system with the outer layer made of a stone-like metal grille while the inner layer is composed of a glass curtain and plain concrete walls. The design uses modern materials to interpret the spiritual characteristics of a traditional indigenous style in a contemporary context. The building volumes on the podiums also have double-layered facades where the outer layer features imitation wood, made of aluminum alloy, to resemble Tang Dynasty's wooden structure and openings on exterior walls. All collaborate to emphasize the connotations of traditional Tang style architecture.

建筑细部 Building details

交响乐观众厅 Symphony hall interior

# 评论
# Review

西安曲江的大唐不夜城贞观文化广场，作为承载盛唐文化的载体，同时能体现出现代文化空间的消费价值，正是通过不断的研究和思考，最终用创新的理念来诠释西安的城市文脉，把盛唐文化用现代的手法和材料来演绎的成果。在这个挑战与创新并存的过程中，我们更加深刻地感受到了城市文化的再造不仅仅是对昔日生活场景的再现，更是对城市文化、空间、细节的理解和诠释，使其既承载历史又是现代的生活中不可缺少的一部分。[1]

郑滢

天华集团品牌公关总监、建筑委员会秘书长

"大唐不夜城文化广场"项目开始于2005年，正值建筑学界探索如何诠释"新唐风"、在秉承传统的同时去实现创新的时期，由此，这个项目中的设计思考具有批判性。

大唐文化广场位于曲江区大唐不夜城的核心，整个大唐不夜城规划要求是复古唐风，当时其他项目普遍是"现代建筑墙身+仿唐坡屋面"的设计手法。项老师认为用地选址在大雁塔旁，这里属于历史敏感地带，他不想造一个仿唐的假古董。而是提出"文化大殿+现代高台"的设计意象，参照佛光寺大殿，还原了一个唐代原汁原味的大屋顶来容纳歌剧院、音乐厅等演艺空间；构建了一个双层表皮的现代基座来布置各类配套空间。最终建成的作品低调谦逊，与大雁塔和谐统一，这种现代与传统的并置，既是对"新唐风建筑"设计的探索，更是对历史街区风貌再现的重新诠释。

程翌

上海翌建建筑规划设计事务所主持建筑师
建筑学博士后

The Great Tang Everbright Town Zhenguan Cultural Center serves as a carrier of the prosperous Tang culture while reflecting the value of a modern cultural space. This effect results from continuous research and innovative ideas to interpret and show the history with modern techniques and materials. Within this challenging and innovative process, we understand that the reconstruction of city culture is not a reproduction of the old. Rather it acts as a new expression of the culture, space and details of the city. It fuses into modern life and it possesses its history.[1]

Zheng Ying

Brand PR Director/ Secretary General of Architecture Committee, Tianhua Group

The Great Tang Everbright Town Cultural Center project began in 2005, the same year when the architecture academia was exploring a reinterpretation of the "Neo-Tang-Dynasty Style" that would be both innovative yet adhering to the tradition. Therefore, the design process of this project bore great significance.

The Cultural Center is located at the core of the Great Tang Everbright Town in Qujiang District, which was supposed to be built completely in Tang style according to the planning scheme. At that time, a common practice elsewhere was fitting a Tang style roof onto a modern building. However, Mr. Xiang felt strongly that it would be inappropriate to build a fake antique structure on a historically sensitive site adjacent to the revered Dayan Pagoda. Instead, he developed the design strategy of creating a cultural hall and modern high platform which echoed the essence of the Foguang Temple: a large authentic Tang style roof accommodating a performing arts spaces such as an opera house and concert hall. A modern podium with a double facade contains supporting spaces. Finally, the design represents a sense of humbleness that coexist harmoniously with the Dayan Pagoda. This juxtaposition of modernity and tradition is not only an exploration of "Neo-Tang-Dynasty Style", but also a reinterpretation of the historical district.

Cheng Yi

Chief Architect, Shanghai Yijian Planning and Architectural Design Co., Ltd.
Postdoctor of Architecture

1 郑滢：《盛唐文化的现代演绎——西安曲江大唐不夜城贞观文化广场规划与建筑设计》，载温晓诣主编《理想空间60：文化再造》，同济大学出版社，2013年
Zheng Ying, "Cultural Renovation, modern interpretation of Tang culture," (in Chinese), Ideal Space 60: Cultural Renovation, Dec.(2013).

正立面实景 Front view of the theatre

2005 / 2017

# 陕西大剧院

作为贞观文化广场建筑群的收官之作，陕西大剧院于2017年正式启幕，是大唐不夜城片区规模最大、硬件水准最优、科技含量最高的单体建筑，它与西安音乐厅共同成为中西部地区最大的国际化表演艺术中心。

从2005年到2017年的十二年，建成后的大唐不夜城贞观文化广场已成为西安市民休闲生活的重要场所，除大剧院外的三组文化建筑在2009年投入使用后广受好评，在城市文化传播上发挥了重要作用。陕西大剧院不仅是大唐不夜城片区浓墨重彩的收笔，也开启了未来城市文化的新篇章。其总建筑面积52 324平方米，设1957座歌剧厅与525座戏剧厅，造型延续了大唐不夜城“文化大殿＋现代高台”的建筑群风格，以大小屋顶的组合及双层基座的造型来满足室内空间使用的需要。

然而十多年的时代发展，技术领域的快速革新、艺术审美趋势的变化，无疑为这座建筑的空间设计提出了新挑战。外立面造型及规整轮廓的限制，抑制了内部公共空间的自由生长，在延续原有公共空间格局和主要挑空区域的基础上，项秉仁及团队在限制中积极寻求突破，为室内赋予了一系列动态曲线，使得不同区域自由流动与组合。灵活可变的艺术展览空间，也为未来提供更多可能性。

在风格手法上，“唐形”“唐意”“唐色”三大元素作为“唐风”的现代表现演绎形式，贯穿各个重要的内部区域，创造出戏剧性的空间感受：

唐形：入口大厅挑空，四周连续的曲面自由而柔和，如霓裳羽衣般灵动。大厅通往二层的螺旋楼梯，与一层吊顶侧面自然形成连续的曲面，打破了矩形门厅空间沉闷、刻板的格局，形成剧院空间的铺陈和引导。大厅与侧厅由于高差变化，吊顶以梯田式流线型曲面处理，不同空间自然连续而整体，让观众行走于如歌舞画卷般的曲面空间中，浸染戏剧艺术的魅力。

唐意：“莲”在东方文化中代表圣洁与宁静。陕西大剧院最重要的歌剧厅设计中，室内天面、墙体和侧向栏板成为星空、莲花、水波的载体，隐喻“星空盛莲”的华丽景象。浓郁的中国红在众星璀璨的幕布下，赋予了整个空间浓重的仪式感，隆重而优雅。

唐色：唐三彩是唐文化里程碑式的艺术色彩形式，一层公共大厅的对景墙面取其白色，结合艺术家的创作，用4000余块不同肌理的陶块拼贴而成，创造了独一无二的墙面表皮，带着“高山流水遇知音”的美好寓意。而戏剧厅大面积取其黑色，暗示空间的稳定性，演绎未来感。黑幕如同吞纳一切五光十色的容器，绿釉般的座椅，暗示跳跃性和节奏感，与整体氛围拼贴出一种超现实的意境。

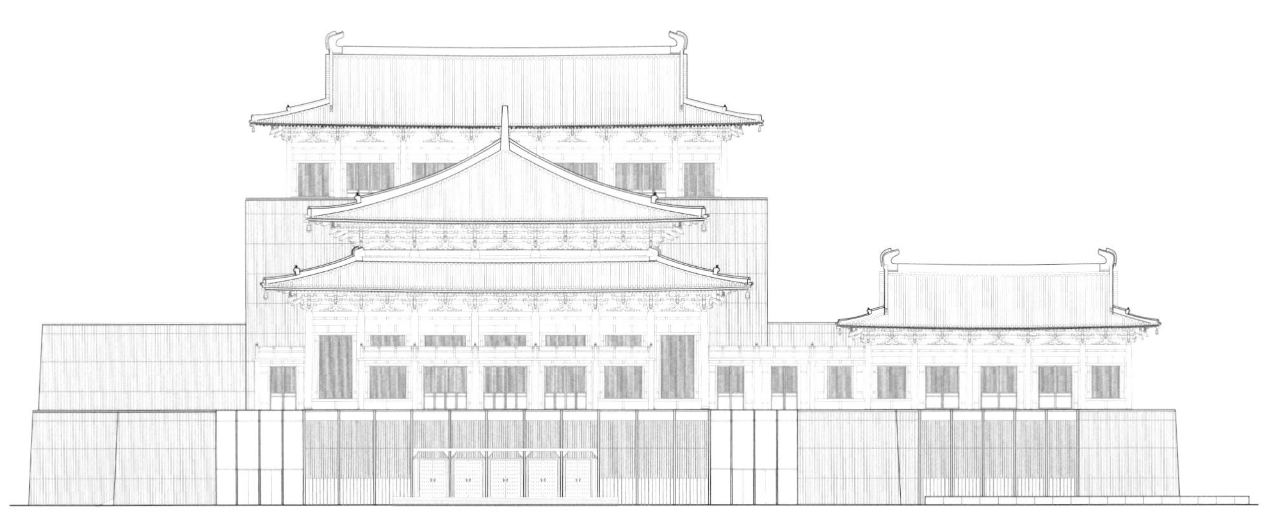
正立面图 Elevation

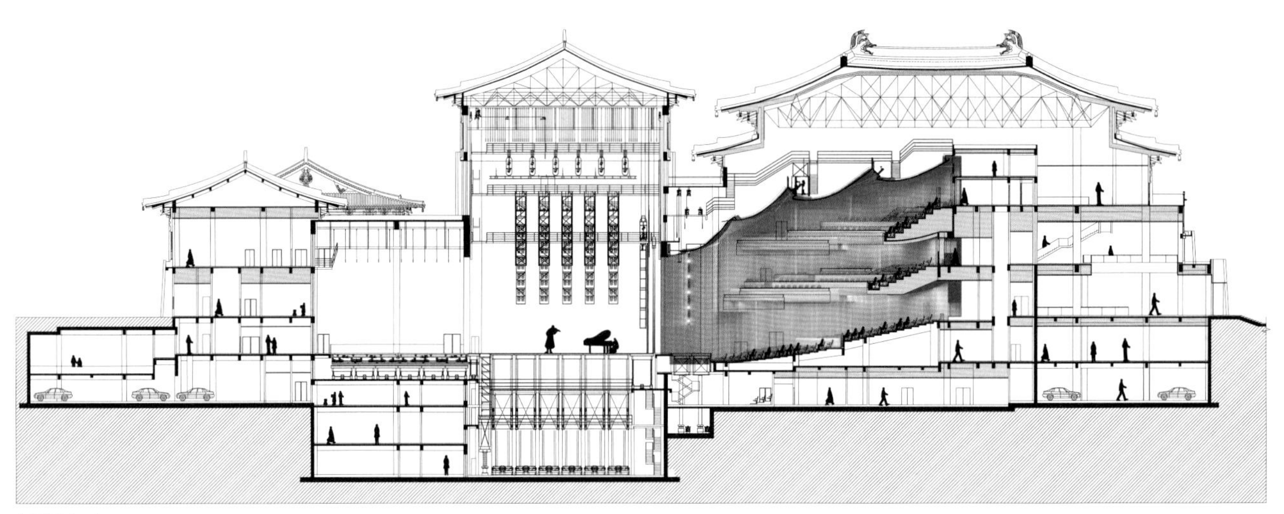
剖面图 Section

剧院是一座城市的文化缩影，承载着城市的历史，滋养着城市的气韵。透过陕西大剧院这件历经十二年的设计作品，可以看出项秉仁对这座城市背后的历史源流与文化传承的坚持，在文脉中不断寻找那些产生共鸣的元素，在历史和现代之间展开对话。这是他对文化基因的阐释与回应，也是留给这座城市里程碑式的一笔。

Officially opened in 2017, Shaanxi Opera House is the final project of Zhenguan Cultural Square complex, also the largest, best equipped high-tech building in Great Tang Mall area. Together with Xi'an Concert Hall, it has become the largest international performing arts center in the mid-western region of the country.

Zhenguan Cultural Square in Great Tang Mall area has grown into a popular recreational venue for Xi'an citizens during its 12-year completion period from 2005 to 2017. The other three cultural buildings in the complex excluding the Opera House have been widely acclaimed since they were put into use in 2009, and now play an essential role in urban cultural communication. The Opera House helps open a new chapter for this communication now that it finally joins the family. With a total construction area of 52,324 square meters, it accommodates a 1,957-seat Opera Hall and a 525-seat Theatre Hall. In appearance it continues the unified form of "a traditional roof + a modern high platform" in Great Tang Mall complex, using the combination of large and small roofs and a double-layer base to create indoor spaces.

However, rapid development in technologies as well as in the field

大剧院室内 Opera house interior

of aesthetics during the twelve-year span inevitably brings new challenges to the space design of this building. The shape and regular contour of the facades restrain any chance of free generation of interior public spaces. Xiang and his team respond to these constraints by creating a series of dynamic curves that enable free flow and interaction between different areas, while maintaining the original public space layout and the main double-height area. A versatile art exhibition space is conceived to provide more possibilities for future use.

In terms of style and techniques, three elements—"Tang shape", "Tang implication" and "Tang color"—are used as contemporary interpretations of Tang Dynasty style in all important interior parts, creating a dramatic spatial experience.

Tang shape: The continuously curved surfaces around the double-height entrance hall resonate with the free flow and elegance of Tang style clothes. The spiral staircase leading to the second floor constitutes a naturally continuous curved surface together with the side of the first-floor ceiling, breaking up the otherwise stereotyped layout of a rectangular foyer space. This sets the tune for the entire space. Due to the height difference between the entrance hall and the side hall, the ceilings are designed as terraced streamlined curved surfaces. Here different spaces are flowing into each other like in a traditional scroll, inviting visitors to immerse themselves in the charm of drama art.

Tang implication: The imagery of a lotus is borrowed for the highlight in the Opera House—the Opera Hall, as lotus embodies holiness

室内大厅 Lobby interior

and tranquility in oriental culture. The indoor sky, walls and side railings signify the starry sky, lotus and water waves, together they depict a spectacular scene of "flourishing lotus under the starry sky". The use of rich Chinese red gives the whole space a strong sense of magnificence and elegance against the starry backdrop.

Tang color: The palette of interior space has drawn inspiration from Tang tri-colored glazed pottery, a significant artistic form of Tang culture. White is used on the uniquely collaged wall facing the entrance on the first floor, consisting of more than 4,000 pieces of ceramics with different textures. It bears the wishful meaning of "meeting a kindred spirit amongst the lofty mountains and flowing waters", one of the most representative imageries in Chinese literal classics. Whereas in the Theatre Hall, a large amount of black is used to bring a sense of stability, but also to communicate a futuristic feeling. The black drop is a vessel that devours all colors; the green glazed-like seats inject dynamics and rhythms, creating an almost surreal atmosphere within the space.

Theaters are the epitomes of a city's cultural life. They carry the city's history and help build its spirit. The realization of Shaanxi Opera House after twelve year's work is evidence of Xiang Bingren's insistence on the historical origin and cultural inheritance of the city. He tirelessly sought for the elements that could resonate in the context of history, so as to start a dialogue between past and present. This is his reading of and response to the cultural gene, and his homage to the city of Xi'an.

## 评论
## Review

陕西大剧院作为“西安大唐文化广场”的收官之作，从2005年开始设计，到2017年建成，时间跨度为12年。今天在曲江新区的主轴线上，大唐文化广场这组文化建筑已经成为不可替代的城市文化核心。虽然建筑形态上，陕西大剧院与其他三个场馆一脉相乘，延续“传统大屋顶”+“现代高台”的造型方式，但其内部的空间已经悄然发生了改变。

陕西大剧院在设计之初，也经历了规模的多次调整，从最初的1500座，调整到1800座，直到最终的2000座。在用地受限、造型受制于传统大屋顶的条件下，这个规模不是容易做到的。而设计后期，基于运营的需要又增加了一个500座的多功能厅和500平方米的艺术展厅。不过换个角度思考，任务书升级虽然增加了设计难度，但也可以让陕西大剧院在未来承担更多的城市文化功能。

2000年以后，中国迎来了大剧院建设的窗口期，省级、市级的大剧院就有30多个。这种快速建设难免和后期运营脱节，不少剧院从选址到内部功能配置都没考虑到后期运营的需要，建成后成为没有演出的剧院，“没有金鱼的金鱼缸”。而陕西大剧院的建设，让我们看到国内的剧场设计思维的一种转变。从追求大而全的形象工程、文化地标，转向更加务实和落地的运营思维。这正是大剧院的设计和建设本应寻求的目标。

程翌
上海翌建建筑规划设计事务所主持建筑师
建筑学博士后

Shaanxi Opera House is the final work of Xi'an Great Tang Cultural Square, a complex that took twelve years from 2005 to 2017 to design and construct. Sitting on the main axis of Qujiang New District, this cluster of cultural buildings has become an irreplaceable core of urban culture. In terms of architectural form, the Opera House adopts the same model of “traditional roof + modern high platform” as the other three venues finished previously, yet inside it has upgraded in line with latest disciplinary developments.

The scale of the Opera House went through several changes at the beginning, from the initial 1,500 seats to 1,800 seats, until finally 2,000 seats. Not an easy task to give the land constraint and formal requirements of “traditional large roof”. To make it even harder, later in the design process, a 500-seat functional hall and a 500-square-meter exhibition hall were added in order to facilitate future operation. Despite these increased difficulties, the silver lining here is the Opera House will take up more cultural functions and play a more important part in the city in the future.

Since the beginning of this millennium, China has been embracing a window period for the construction of theater buildings. Until now there are more than thirty grand theatres at the provincial or municipal levels nation-wide. Construction at such speed brings the risk of lacking integrated strategy, not fully considering the operation side after these buildings are put into use. It’s not news that many newly finished theaters don’t even hold any performances. They become “goldfish tanks without any fish” because they didn’t think through issues like site choices and internal facilities and operations in the first place. Fortunately, Shaanxi Opera House tells us it’s not always the case. It shows a shift in mindset when it comes to theater design, a shift from image engineering and cultural landmarks to a more pragmatic and on-earth operation thinking. And that should be the goal for any “grand Theatre” to pursue.

Cheng Yi
Chief Architect, Shanghai Yijian Planning and Architectural Design Co., Ltd.
Postdoctor of Architecture

建筑组群鸟瞰 Bird's-eyes view of the site

2008 / 2012

# 宁波文化广场

宁波文化广场是与东部新城行政中心几乎同时开发的另一个宁波市政府“十一五”规划重点建设项目。基地位于宁波东部新城中央走廊，占地22公顷（包括两条河道），地上建筑面积约20万平方米，是集文化艺术、科普展示、教育培训、健身养生、服务娱乐等多功能于一体的城市功能片区和文化娱乐目的地。其中包括大剧院、科学探索中心、CGV影城、艺术培训中心、少儿体验中心、乐高活动中心、游泳健身中心、朗豪酒店八大主力项目和一系列商业文化设施。

文化广场的设计是一次多维度的、从城市尺度到建筑尺度皆具有挑战的实践。它缘起于宁波市东部新城规划局组织的一次国际竞赛。在项秉仁参与这次设计竞标之前，业主方已经组织过两轮设计方案征集，却并未选出令他们满意的设计方案。在仔细踏勘项目基地，充分了解项目背景和业主诉求之后，项秉仁首先面临着一个抉择——二十多万平方米的地上建筑体量是作为一组建筑来处理，还是作为一个城市片区来设计。在当下的建筑实践中，这样的体量对于一座建筑综合体而言并不算太大。但项秉仁却意识到，这是一个塑造城市中心片区的机会，不应局限于简单的建筑设计，也并非只是打造一个以“文化广场”冠名的巨型地标建筑。基于一种新城市主义的理念，项秉仁有意识地在概念方案设计过程中主动加入了城市设计阶段（虽然规划局和业主并不要求如此），使最终优化的城市设计和导则成为下一阶段建筑设计的条件依据。这一做法有利于暂时撇开琐碎细节的建筑设计思考，集中注意力于基地内街道广场等城市空间的营造，特别是城市滨水空间的整体性处理。

文化广场的整体空间分为四个板块。项目中央是以东西向的河道主轴和南北向的河道次轴为骨架的滨水空间体系。在水道汇聚之处设置了纵横的步行联系和突出水面的露天广场和表演区，它们是整个区域城市空间的核心。四个地块的滨河部分和区块内部分别设置了城市广场节点和步行交通体系，它们充分相连，形成了充满活力和可能性的城市空间。东北区块设有一栋1500座规模的大剧院以及一栋CGV影院综合体。两座建筑体量较大，采取大实大虚的手法，是整个文化广场的精神和文化制高点。西北区块主要由儿童滨水娱乐场所以及科技展示中心组成。其建筑体量较

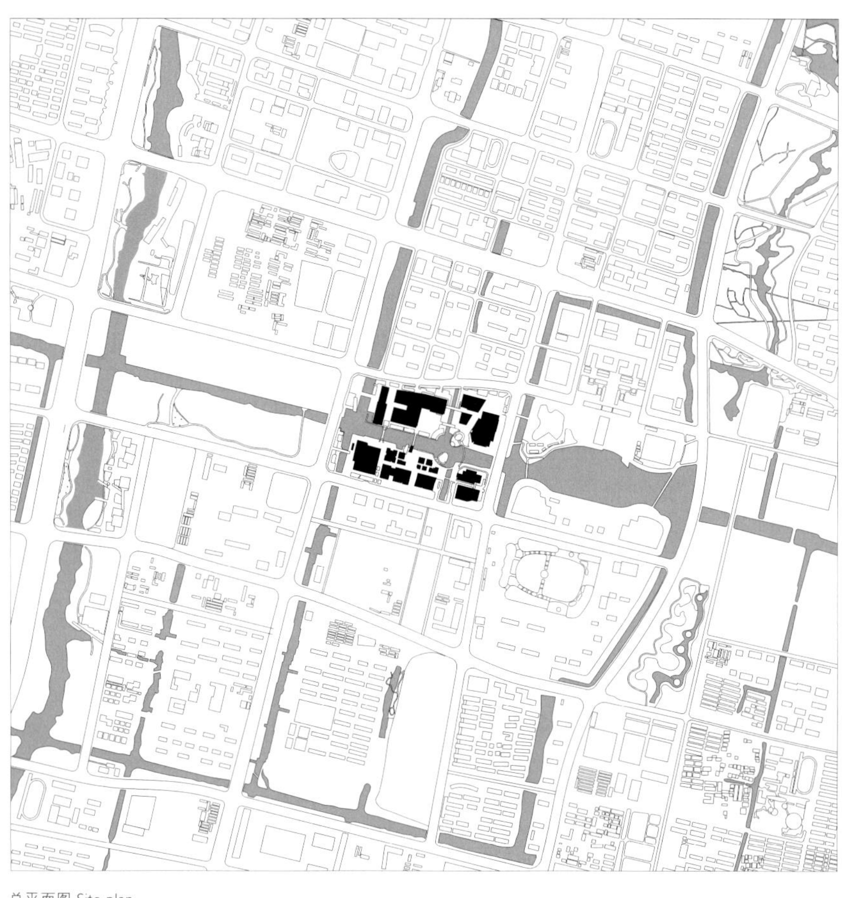

总平面图 Site plan

舒展并结合了巨大的科技馆前广场空间，是整个文化广场最具备大型室外活动和人流汇聚功能的区域。西南区块包含儿童娱乐中心、群众文娱艺术馆及时尚健身馆和滨水商业街。这里建筑尺度相对较小，步道阡陌纵横，是适合步行和休憩的区域，也是文化广场承载商业人气的核心区域。东南区块相对独立地布置了朗豪酒店和艺术展廊，是对项目整体的配套支撑，其酒店塔楼也是文化广场的竖向坐标，是天际线中不可或缺的一笔。

完成了城市设计层面的操作后，设计的另一个特别考虑就是有意识地强调在城市设计导则控制下的建筑物的不同个性，一方面是为了营造一处既统一又有变化的城市环境片段，另一方面也隐喻着城市发展进化的时间性。为了实现这个目标，项秉仁有意让事务所内的建筑师分别负责设计各个不同地块的建筑物，通过百花齐放、更具偶然性的建筑组群设计取代以往地块内完全整齐划一的建筑语言表达。于是我们今天看到的文化广场就以这样的方式出现了——建筑师努力塑造差异性，甚至希望来到这里的体验者会不经意地认为这些建筑是出自不同的建筑师之手的。每一栋建筑的表皮和材料都各不相同，这当然给设计和施工带了一些难度，但却形成城市片区的形象有别于通常的建筑师大笔一挥而造就的划一形象。它模拟了生长性，是对东京的表参道商业建筑群，或者阿姆斯特丹的港区更新或者宁波传统街区的致敬，是在严谨

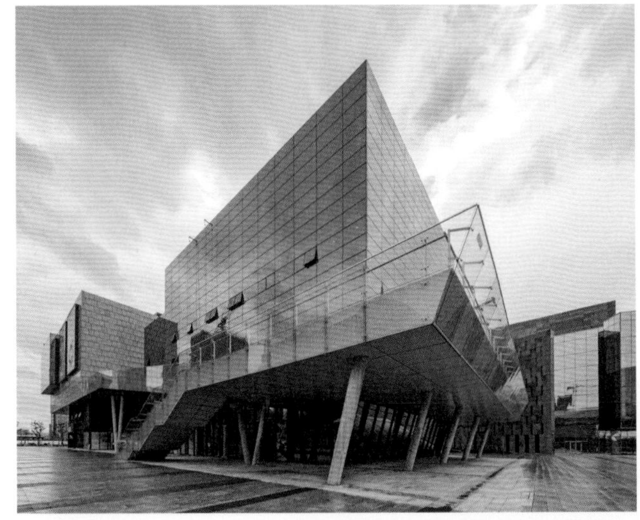

实景照片 Building images

的城市设计控制下生发出多样性和个性表达的一次尝试。

文化广场的设计当然也是一次关于建筑美学和建筑材料的试验。每一栋单体建筑的表现手法都不尽相同，由此产生了很多的材料和构造节点。大剧院和影城的大面积玻璃幕墙与石材幕墙的设置、科技馆的各种金属材料的运用、儿童滨水娱乐馆的玻璃幕墙结合水平遮阳系统的设计、朗豪酒店的平面化的玻璃幕墙与石材幕墙的拼贴，商业街建筑群的穿孔金属板、陶管、陶板、玻璃装饰幕墙、玻璃肋装饰百叶等众多的设计节点在设计师与合作单位的努力下一一得以实现，最终呈现出一组具有高度城市学的统一性与建筑学的随机性，具有理性却又充满活力的建筑组群。

The Ningbo Cultural Plaza was another key construction project for the "11th Five-Year Plan" of the Ningbo Municipal Government and was developed at the same time as the eastern New Town Administrative Region. The site is located on the central corridor of the eastern part of the City. It covers an area of 22 hectares (including two rivers) with a 200,000 square meter projective building area which supports services and educational programs for cultural activities, arts, education & training, fitness & health services, entertainment, and even a science learning center, among other functions. Major construction includes a Grand Theatre, a Science Discovery Center, a CGV Studio, an Arts Training

从万人广场看大剧院和影城 View of the theatre and cinema

Center, a Children's Experience Center, a LEGOs Events Center, a Swimming Fitness Center, a Langham Place Hotel in addition to eight other pieces as well as a series of commercial and cultural facilities.

The design of the Cultural Plaza is a cross-scale practice that is challenging on both urban and architectural levels. It originates from an international competition organized by the Ningbo Metro Planning Bureau. Before we participated in the design bidding for the project, the client had previously organized two rounds of requests for proposals which resulted in no satisfying submissions. After carefully examining the project site and fully understanding the background of the project and the client's vision, Xiang Bingren was faced with a choice: whether to treat the over 200,000 square meters of above-ground building volume as a group of buildings or as a metropolitan area to be designed. In current architectural practice, such a volume is not too massive for a building complex. However, Bingren realized that this was an opportunity to shape the center of a city and should not be limited to a basic architectural design. He estimated that the project was more than simply building a giant landmark falsely bearing the title of a "Cultural Plaza." Based on concepts typical of New-Urbanism, Xiang Bingren consciously placed an urban design stage (although not required by the Planning Bureau nor the client) before the conceptual architectural design and made the finalized urban design a guide for the next stage of the architectural design. This approach helped to temporarily avoid the trivial thinking of architectural designers and focus on the creation of urban spaces such as the streets and plazas within the site, especially the integration of the urban waterfront.

The Ningbo Cultural Plaza is divided into four sections The waterfront is central to the design and stretches along an east-west river as the main axis and a north-south river as its sub-axis. There are vertical and

horizontal walking links and urban plazas in addition to performance areas at the converging points of the waterways, forming the core of an urban space for the entire area. The plaza's and pedestrian circulation systems have been set up in both the riverside areas and interiors of the four sections. All are fully connected to form an urban space full of energy and possibilities. The 1500-seat Grand Theatre and a CGV theater complex in the northeast sector apply the strategy of using huge spatial voids to balance the large solid volumes created by the surrounding buildings. These features serve as the spiritual and cultural high ground of the entire cultural plaza. The north-western district hosts a child's waterfront playground and a technology exhibition center. The buildings stretch horizontally and include large square spaces in the front. It is the largest venue in the area for large-scale outdoor activities and can accommodate many people. The southwest section is arranged as a child's entertainment area, public entertainment center, gym and waterfront commercial street. Although relatively small, the spread of the area includes dense trails arranged for walking and rest and attracts crowds to the project's core business district. The southeast block is relatively independent. It is arranged with Langham Hotel and an art gallery, providing support services for the cultural plaza project. The hotel tower also anchors the project vertically while serving as an integral part of the city skyline.

Following the completion of the urban design phase, another special consideration of the design was to consciously emphasize the different personalities of the buildings under the control of the urban design guidelines. The architects needed to create a unified yet diverse urban fabric while also implying the phasing of construction. To achieve this goal, Xiang Bingren deliberately asked different architects in the firm to take charge of the architectural design of different plots instead

A-1
B-4
B-2
B-3
A-3
A-3
B-1
B-1
E-1
E-5
E-4
G-2
G-3
G-5
G-6
G-1
G-1
G-4
G-1
G-1
E-1
E-2
E-3
F-2
F-1
F-3
F-4
F-5
F-6
F-7
F-8
F-9
F-10
F-11
F-10

A 滨水娱乐
A-1 门厅
A-2 甜品店
A-3 文化休闲

B 科技馆
B-1 科技展示
B-2 咖啡吧
B-3 临展前厅
B-4 科技馆临时展厅
B-5 科技馆前厅
B-6 纪念品销售
B-7 存包处
B-8 信息中心
B-9 影院门厅
B-10 中心网站
B-11 电梯厅

C 影院
C-1 零售
C-2 消防控制中心
C-3 商业概念店
C-4 商业概念店
C-5 综合性概念店
C-6 入口大厅
C-7 咖啡
C-8 餐饮
C-9 入口大厅
C-10 主题商业

D 多功能剧场
D-1 前厅
D-2 休息厅
D-3 静压箱
D-4 精品餐饮门店
D-5 贵宾室
D-6 贵宾门厅
D-7 侧台台仓
D-8 后台台仓
D-9 钢琴房
D-10 中化妆间
D-11 大化妆间
D-12 门厅

E 儿童娱乐
E-1 探索乐园门厅
E-2 科学迪士尼门厅
E-3 儿童用品零售
E-4 儿童职场体验入口大厅
E-5 休闲商业

F 群娱中心
F-1 群娱门厅
F-2 咖啡
F-3 消控中心
F-4 书局
F-5 创意照
F-6 礼品
F-7 群娱辅助门厅
F-8 琴行
F-9 花艺
F-10 创意零售
F-11 画廊

G 商业
G-1 休闲商业
G-2 中餐厅
G-3 厨房
G-4 特色酒吧
G-5 特色餐厅
G-6 厨房
G-7 艺术酒吧
G-8 特色酒吧
G-9 创意零售

H 健身中心
H-1 运动品牌零售
H-2 休闲商业

I 文化交流会所
I-1 酒店主要入口
I-2 大堂
I-3 西餐厅
I-4 厨房
I-5 后勤办公
I-6 员工入口
I-7 消防控制中心
I-8 酒店服务

J 艺术沙龙
J-1 艺术沙龙
J-2 展厅
J-3 庭院

总一层平面图
First floor plan

实景照片 Building images

建筑细节 Site view

of following the common practice of redeploying the same design in different areas. This resulted in a diversified architectural group. So, the Cultural Plaza we see today appears in such a way that architects strive to create differences and even hope that those who come here will inadvertently realize that these buildings are from different architects. The design and materials of each building facade are different, resulting in some difficulties in the design and construction. But the trade off is worth it as the approach brings a rich and varied image to the urban area and simulates the natural growth of the city. Such an approach was inspired by methods used by Tokyo's Omotesando commercial complex and the renovation of the port area of Amsterdam. It also paid tribute to the traditional district of Ningbo. The design was an attempt to generate diversity and individuality under rigorous urban design control.

Of course, the design of the Cultural Square is also a testing ground for architectural aesthetics and building materials. Every single building has a different appearance, resulting in many different material applications and detailed construction designs. Several examples: The large-scale application of the glass curtain wall and stone curtain wall in the Grand Theatre and the Film & TV City; the use of various metals in the Science and Technology Museum; the design of the glass curtain wall at the children's waterfront entertainment hall combined with the horizontal shading system; the planarization of the Langham Hotel's Glass curtain wall and stone curtain wall collage; the Commercial Street' perforated metal plate, ceramic tubes, ceramic panels, as well as the glass decorative curtain wall, glass rib decorative blinds, among others. Thanks to the efforts of the designers and cooperators, numerous detailed designs have been realized, resulting in a series of buildings that are both highly unified in the city and with each possessing its own uniqueness, together representing rationality and dynamism at the same time.

## 评论
## Review

在设计之初我们为这组建筑和空间取名为“文化的港湾”，在描述其精彩的滨水空间的同时，更是对其文化汇聚和人气承载的期待。我们希望建筑群体在形态、色彩、材料和谐统一的前提下，形成具有独特个性的文化商业聚集空间，体现当代宁波“海纳百川，兼容并蓄”的开放性都市气质，呈现意趣盎然的空间形态序列；同时结合宁波城市的“水”文化和“院落”民居特色，呈现较为鲜明的建筑地域特色。

滕露莹
上海秉仁建筑师事务所合伙人、副总建筑师

中国的城市空间新形象通常难以出现在重金加持、权力集中的一二线城市，也不容易出现在人口流失、增长乏力的中小城镇，但是往往出现在具有历史沉淀、经历了现代主义洗礼、努力塑造自我认同的2.5线城市，比如宁波这样的城市。如果说宁波老城区的“天一广场”遵循的是一种中心化、聚合式的城市公共空间传统，那么东部新城的“文化广场”创造的则是一种去中心的、弥散的体验式的城市公共空间新模式。这种模式强调沟通、包容、平等、共享，摒弃中心主导或者零和博弈式的空间运作，呼唤民主和定制，面向创新与未来。既不凭借穷形尽相的标志建筑来夺人眼球，也不依赖夸张尺度的城市广场来吸纳人流，而是以一种平和自信的方式营造空间、排布建筑、刻画细节，宁波文化广场的空间逻辑洒脱而节制，简约又生动，带给人的感受是：适宜的尺度、干净的造型、精致的界面、多变的体量。

“从心所欲不逾矩”是建筑师项秉仁先生在宁波文化广场设计中流露出的气度和姿态。正如冯纪忠先生所言：“我受自然的孕育而不要众人瞩目于我。”于是我们有幸看到今天新城核心区的宁波文化广场，没有一般商业街区的刺激与媚俗，也没有普通文化建筑的恣肆或清高，而是“和而不同”地传递出设计者对文化建筑内涵的理解和场地街区气氛的把握。

程雪松
上海大学上海美术学院院长助理、设计系副系主任、教授、博士生导师

At the beginning of the design, we named this group of buildings as a “harbor of culture”. It is simultaneously a wonderful waterfront space and serves to gather culture while accommodating various activities. Under the harmonious unity of form, colors and materials, we hope to create a unique space gathering cultural and commercial activities. At the same time, we hope the design can reflect the contemporary and open urban temperament of Ningbo. Moreover, with the modern architectural language and the design of different building facades, we intended to construct an intriguing sequence of spaces that combine with the “water” culture of Ningbo city and the characteristics of the “courtyard” residential buildings to represent the distinctive regional architectural characteristics. Today, five years after the completion of the project, the vision for the harbor has been realized. With the increase of popularity and the maturity of independent commercial operations, the Cultural Plaza has only improved in its mission to aggregate people and activities. Those urban spaces and diverse architectural forms we designed, following New-Urbanism ideal, also demonstrated its flexibility and ability to accommodate multiple functions throughout the project's operation. It has been in the urban context for a long time and has slowly become familiar and enjoyed by more and more Ningbo citizens as a symbol of the city itself.

Teng Luying
Partner/ Deputy Chief Archicect, DDB Architects Shanghai

Innovative uses of urban spatial images are a rare sight in first- and second-tier cities where power and capital are highly concentrated. Even in small and medium-sized towns they are difficult to find given the frequent population loss and poor prospects. All of this seems to push 2.5 tier cities, such as historical Ningbo, through a modernist phase where they strive to reshape the identities through architectural sites. If the “Tianyi Square” in the old city of Ningbo follows a monocentric tradition of urban public spaces, then the “Cultural Plaza” of the Eastern New Town creates a decentralized, experiential model. This model emphasizes communication, tolerance, equality, and sharing, while abandoning the center-led or zero-sum game of spatial operation. In short, it calls for democracy and customization. Also, this model is oriented towards innovation and the future, relying on neither flashy landmarks to capture attention nor urban squares with an exaggerated scale to attract people. Rather, it seeks but to create space, arrange architecture, and portray details in a peaceful and confident way. The Ningbo Cultural Plaza features a familiar, human scale, clean shapes, exquisite facades, and varied volumes. It follows a spatial logic that is free and easy, simple and vivid.

"Do nothing more than you want" was the attitude and posture of Mr. Xiang Bingren while designing the Ningbo Cultural Plaza. As Mr. Feng Jizhong once said, "I am nurtured by nature and I don't want to be the center of attention.” Therefore, we are fortunate to see the Ningbo Cultural Plaza as the urban core of a new city today, which avoids being overly stimulative and kitschy like ordinary commercial districts. It is also not super inaccessible and out-lying like ordinary cultural buildings, but rather harmoniously yet firmly expresses the designer's own understanding of cultural architecture and the atmosphere of the site.

Cheng Xuesong
Dean Assistant/ Deputy Director of Design Department/ Professor/ Doctoral Supervisor, Shanghai Academy of Fine Arts, Shanghai University

鸟瞰日景航拍 Bird's-eyes view

# Jinhua Science & Culture Complex



2015 / 2019

# 金华科技文化中心

2015年项秉仁和他的设计团队接受浙江金华市的邀请，通过竞标获得了该市拟建的科技文化中心的设计任务。项目基地位于金华核心区城市发展轴线的重要区域三江口多湖中央商务区内，与婺剧院遥相呼应。整体建筑组群包括城市展示馆与展览中心、市民活动中心、青少年培训中心和主题科技馆四大文化场馆及一系列相关的辅助城市功能。

科技文化中心的建筑组群设计源起于项秉仁对已有的城市设计和周边环境的尊重。通过对城市设计中“智慧之树”理念的延续，结合项目的使用要求，运用理性的结构柱网和方正的建筑轮廓去构建高效通用的建筑组群模块——“智慧果实”。各个组群体块在经过转动、变形和错动后呈现出互动与张力，实现了建筑与城市的相互渗透和空间共享。同时，建筑组群界面的限定，强化了婺剧院的入口轴线，在不失个性的前提下，烘托了婺剧院的主体地位，并与之形成统一和谐的整体。

与此同时，项秉仁希望在三江口岸赋予新建的公共建筑一个“城市甲板”的理想，使其成为实现“多馆合一”的公共功能基座，同时包裹住场地内的交通、货运和停车通道，释放更多的空间给城市及景观。台阶和坡道使甲板成为广场空间的延续，在鼓励人们活动的同时，也具有承载各种展览、室外活动、台阶剧场等主题活动的可能性，就像一座巨大的城市会客厅。

立面设计以地域文化和人文情怀为脉络，试图在金华传统文化中寻找“文化之光”。截取东阳具有代表性的金华竹编工艺作为立面表皮设计元素，依据各馆的不同功能属性，通过竹编的变化和统一，构思顺应每一个空间采光、通风及氛围营造需求的不同虚实、疏密的立面表皮，“五彩编织”般构筑成一组“高度融合，合而不同”的文化建筑表情。

建筑室内空间被赋予“艺术之树”的理念，经由室外的“智慧之树”的蔓延，串接起室内的诸多公共空间，向下绵延至地下公共配套功能及下沉庭院，向上串接各馆的中庭空间直至城市甲板空间，打造一个立体化的艺术主题空间。中庭空间、室外的城市甲板和远处的江景得以相互交织在一起。

建成后的金华科技文化中心，就像一个充满活力的生命体，她实现了项秉仁在设计过程中所追求的建筑功能的共享、内外空间的共存、组群形象的共生，最终成为一组标志性的城市文化地标。她以活泼、睿智、丰富多元的文化氛围吸引人们、感染人们、打动人们，成为这座城市精彩的城市文化灯塔。

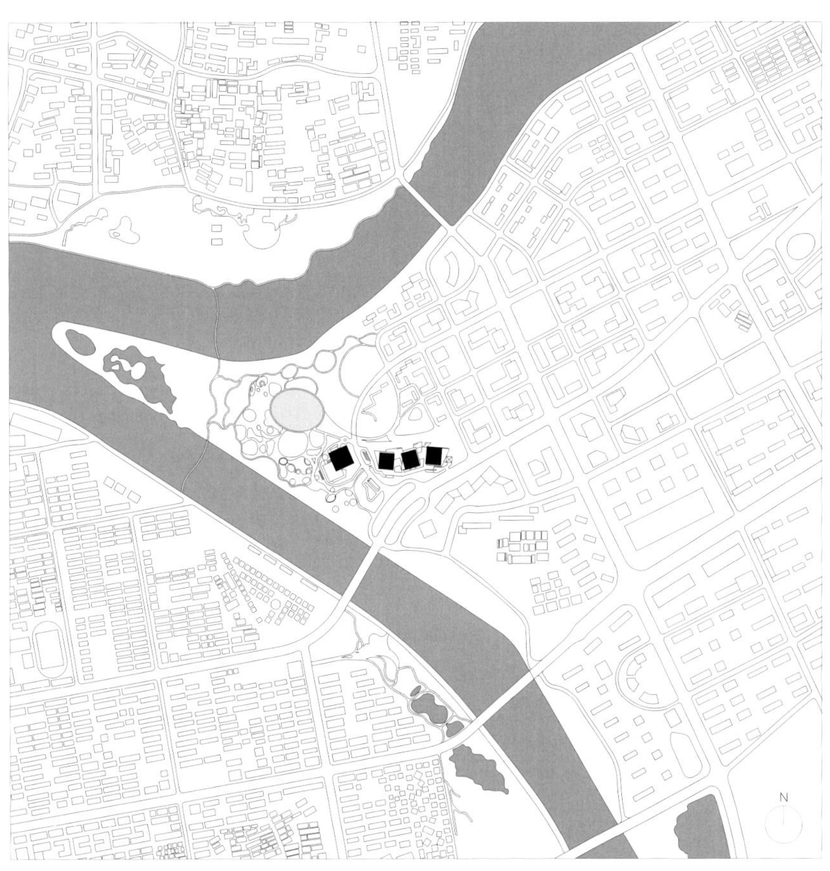

总平面图 Site plan

In 2015, Xiang Bingren and his design team were invited to participate in the bidding for the design of the Science & Technology Cultural Complex in Jinhua, Zhejiang. The project is in the Duohu Central Business District at Sanjiangkou. This is an important area on the development axis of the core area, echoing the Wu-Opera Theatre. The complex includes four major cultural venues such as the City Exhibition Hall and Exhibition Center, the Citizens Activity Center, the Youth Training Center and the Science and Technology Museum, as well as a series of related urban facilities.

The architectural design for the Science and Technology Cultural Center originated from Xiang Bingren's respect for the current urban design and surrounding environment. By continuing the "tree of wisdom " urban design concept, following the functional requirements and adopting the rational structural grid, cuboid volume, the design managed to create an efficient architectural module referred to as the "fruit of wisdom". The group of buildings represents a sense of interaction and tension after rotation, deformation, and misalignment, achieving a pleasant integration and sharing of space between buildings and cities. At the same time, the interface of the group of buildings strengthens the entrance axis of the theater. Without undermining its uniqueness,

< 主题科技馆建筑外观 Building exterior of Science and Technology Museum
> 市民活动中心建筑外观 Building exterior of Citizens Activity Center
∨ 青少年培训中心建筑外观 Building exterior of Youth Training Center

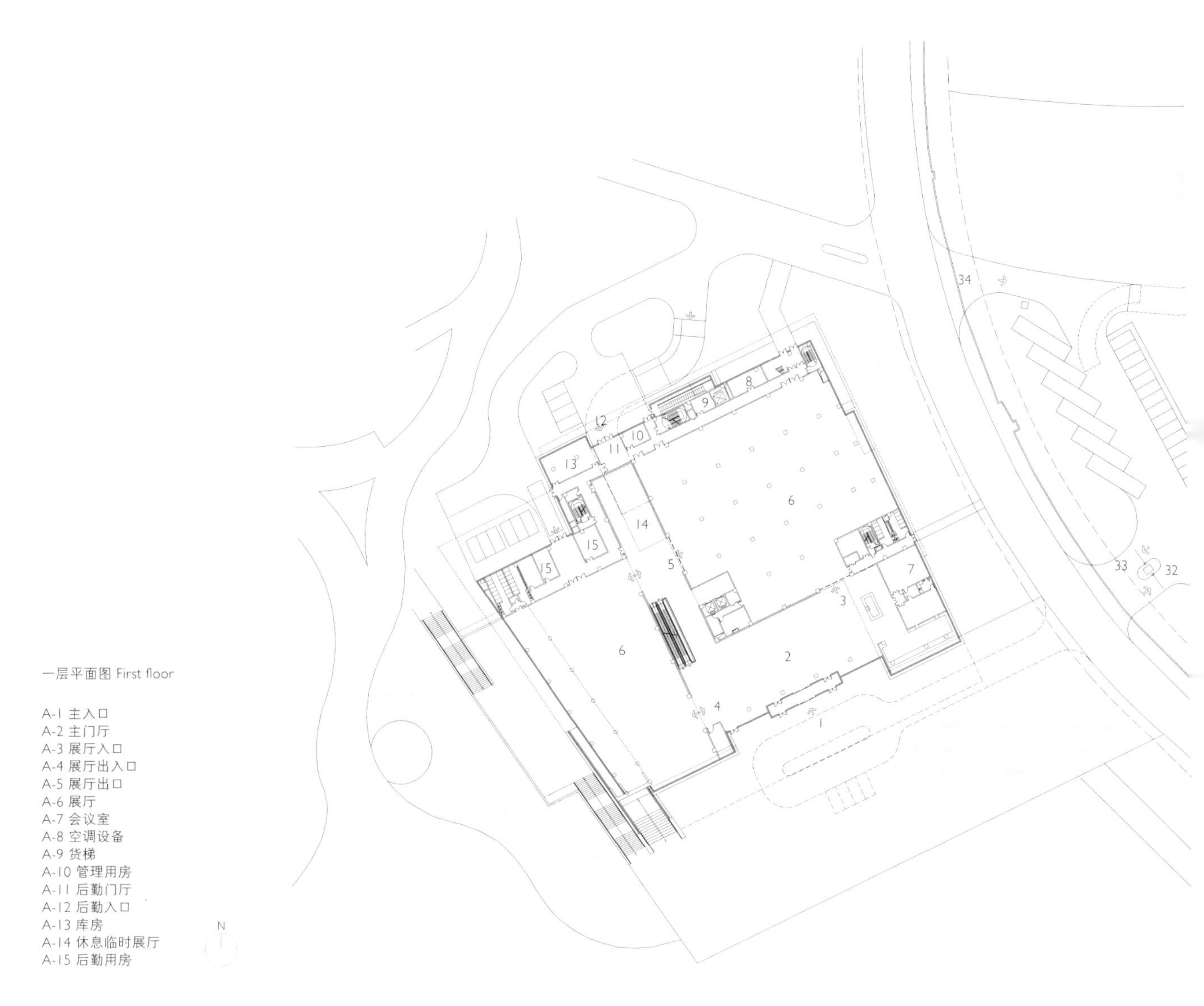

一层平面图 First floor

A-1 主入口
A-2 主门厅
A-3 展厅入口
A-4 展厅出入口
A-5 展厅出口
A-6 展厅
A-7 会议室
A-8 空调设备
A-9 货梯
A-10 管理用房
A-11 后勤门厅
A-12 后勤入口
A-13 库房
A-14 休息临时展厅
A-15 后勤用房

the final layout emphasizes the theater's main position and contributes to the creation of a harmonious whole.

At the same time, Mr. Xiang hoped to create an "urban deck" for the newly-built public buildings at Sanjiangkou Port, turning it into a functional base and holding a variety of public activities while forming a unified whole. The base can help to cover the traffic, freight, and parking lots within the site while also freeing up more space for the city and landscape. Steps and ramps connect the deck to the main plaza, which not only promotes social interaction, but also creates possibilities for various exhibitions, outdoor activities, stage theaters and other thematic activities, just like a huge urban living room.

The facade of the building is based on regional culture and with human proportions and sensibility. It seeks to find the "light of culture" in the traditional culture of Jinhua. The design team assimilates the traditional handicrafts of Jinhua bamboo weaving as the design element of the facade surface. Each hallway varies according to function. The functional differences are represented through variations of bamboo weaving, facades of different "fabric counts" according to each

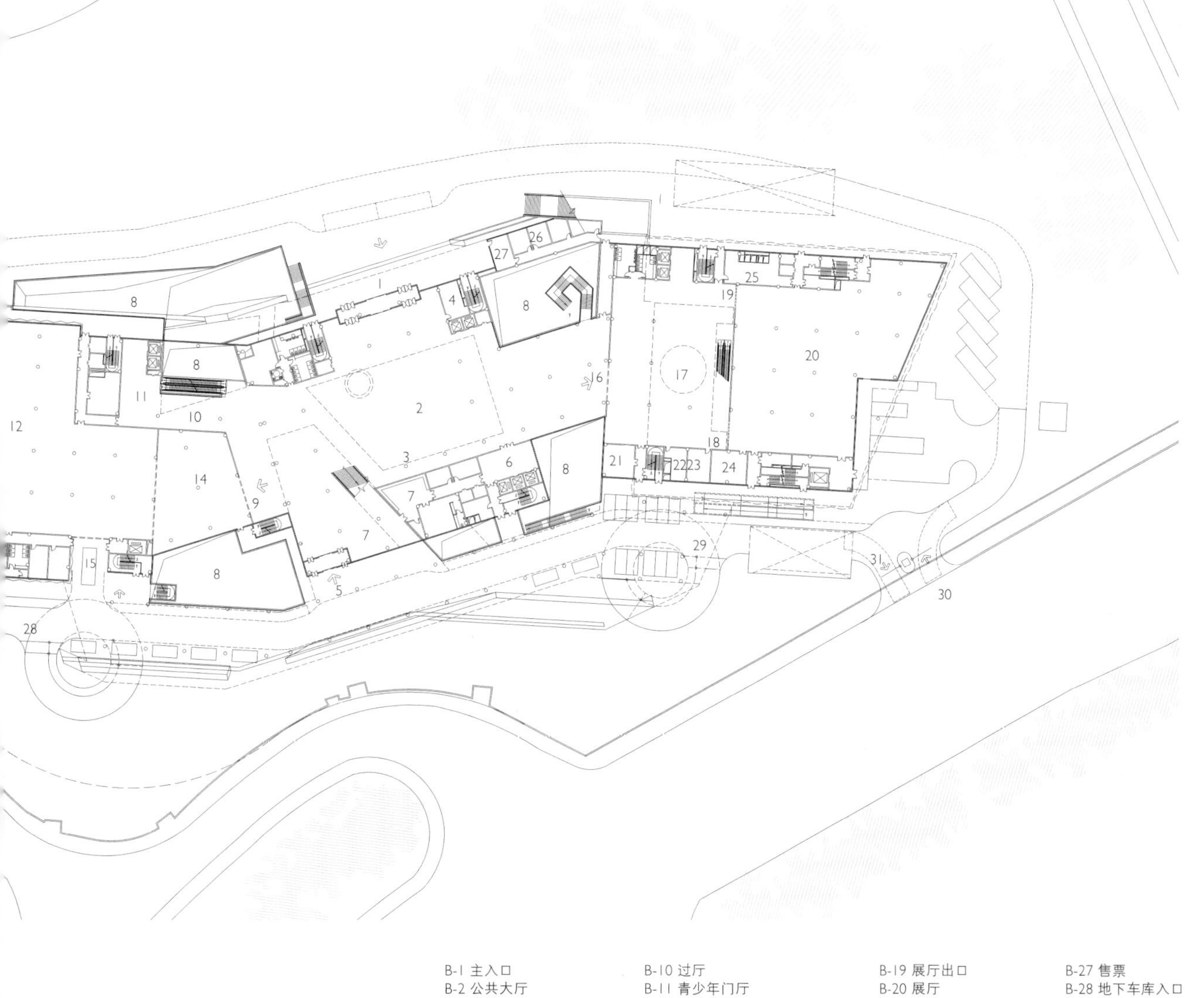

B-1 主入口
B-2 公共大厅
B-3 信息查询
B-4 值班室
B-5 团队入口
B-6 电梯厅
B-7 休息区
B-8 下沉庭院
B-9 儿童体验入口
B-10 过厅
B-11 青少年门厅
B-12 儿童体验
B-13 青少年货梯
B-14 儿童体验前场
B-15 儿童体验后勤入口
B-16 科技馆入口
B-17 中庭区域
B-18 展厅入口
B-19 展厅出口
B-20 展厅
B-21 监控
B-22 排烟
B-23 数据机房
B-24 科普用房
B-25 空调机房
B-26 办公
B-27 售票
B-28 地下车库入口
B-29 地下车库出口
B-30 后勤货运入口
B-31 社会车辆出口
B-32 社会车辆入口
B-33 大巴车出口
B-34 大巴车入口

space's lighting, ventilation and space requirements. "Colorful Weaving" is built into a group of "highly integrated, united but different" cultural building expressions.

The architectural interior space adheres to the "tree of art" concept, connecting the "tree of wisdom" from outside. The "tree of art" connects many indoor public spaces while also extending all the way down to the underground public facilities and sunken courtyard. It also cascades up to the atrium space of each hall and reaches out to the city deck space where it forms a three-dimensional indoor art-themed space. The atrium space, outdoor urban decks, and the riverscapes in the distance are all interconnected.

The completed Jinhua Science & Culture Complex has realized the concept, created by Mr. Xiang at the beginning of the design, that to create a space for sharing, coexisting of outdoor and indoor spaces, and of harmonious symbiosis. It is like a vibrant living body which has become a landmark of urban culture. The project attracts, influences and impresses people with a lively, wise and rich cultural atmosphere, turning this city into a wonderful urban cultural beacon.

甲板之上的东三馆 Building sits on the "city deck"

和而不同的表皮肌理 Skin textures

城市展览馆

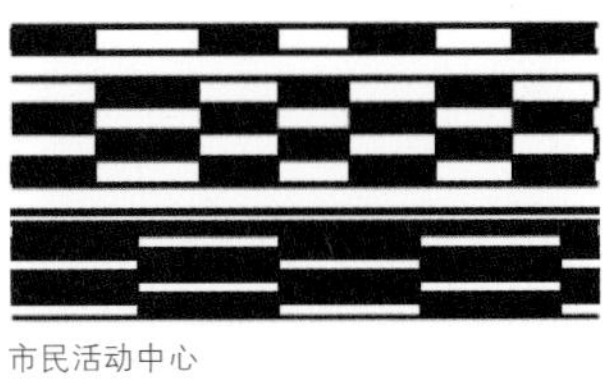

市民活动中心

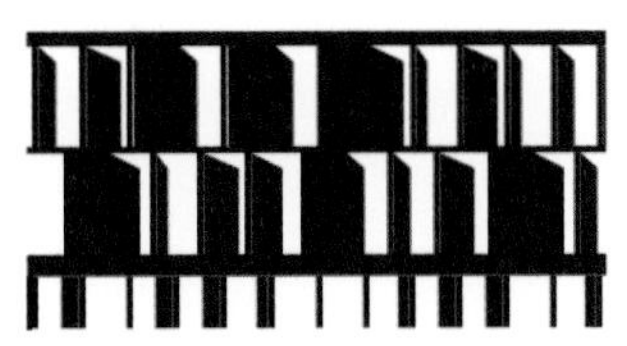

青少年活动中心

主题科技馆

^ 团队入口大厅 Group visitors entrance
∨ 团队入口 Group visitors entrance lobby

青少年活动中心大厅 Hall of Youth Training Center

市民活动中心公共大厅 Public shared space

## 评论
## Review

项老师对金华地域文化特色进行提炼和建筑转译，通过建筑形态的突破创新，从城市设计的宏观角度去协调和优化城市空间，睿智地将功能复杂的多个建筑整合到一个共同的基座上形成了文化地标，并创造了丰富有序的室内共享空间，营造了台阶剧场、观景长廊、亲子平台、绿化庭院等多维的开放空间，塑造了一个充满城市活力的建筑生命体。

《庄子·天道》中提及："不徐不疾，得之于手而应于心"，用这个形容项老师的建筑创作，最恰当不过！

吴丹

同济大学建筑设计研究院（集团）有限公司三院设计总监、高级工程师

Prof. Xiang refines the cultural characteristics of the Jinhua region and redeploys it to architectural design through his groundbreaking innovation of architectural form, coordination and optimization of urban space from urban design perspective. He thoughtfully integrates multiple buildings with complex functions onto a common pedestal, which then became a cultural landmark with a rich and organized public interior. The design has introduced a multi-dimensional space experience including the theater, viewing gallery, parent-child platform, and the green courtyard, creating a building full of life and urban vitality.

Chuang Tzu quotes admiringly in "The Way of Heaven": "If the movements of my hand are neither (too) gentle nor (too) violent, the idea in my mind is realized", which perfectly describes Prof. Xiang's design philosophy.

Wu Dan

Design Director/ Senior Engineer, Tongji Architectural Design (Group) Co., Ltd.

鸟瞰效果图 Bird's-eyes view

2018 /

# 西安碑林博物馆

碑林博物馆位于西安历史发展的重要轴线上，它是中国最大的碑石博物馆，亦是古代经典刻石的渊薮。博物馆在原孔庙基础上扩建而成，建置可追溯到北宋末年，经历代整修，逐渐成为以收藏、研究和陈列碑石、墓志及石刻造像为主的专题性艺术博物馆。然而，这辉煌的传统文化如今却面临展陈空间不足、保存条件差、设施老化等诸多问题。2017年，陕西省政府、西安市政府提出碑林扩建，拟在北区建造新馆，并对碑林所在街区进行同步改造。

项秉仁认为，对内在文化的挖掘是让新老建筑产生关联的触媒。设计希望新的博物馆能以一种更当代、更友善的态度融入城市环境，承载并延续碑林的历史地位，建立一种跨越时间的文化共鸣。

设计强化了碑林博物馆的轴线序列，呼应"一轴两翼"的整体格局，形成历史空间向现代空间的自然过渡，隐喻时间的交错和延绵。建筑从传统的拓碑技法中获得灵感，通过现代的建筑语汇将石刻碑文、纸墨拓印转译为表皮、空间的表达。青石构筑了浑厚稳定的体量，深深浅浅、虚实相间的立面肌理似是拓制力度的不同所造成的。砖墙因应了历史的线索，玻璃窥见了"古""今"的关联。青石、玻璃与砖墙互相照应，好似一个个时间的切片，存续了碑林作为文化空间的真实属性。

地面层设置门厅，中庭覆盖在一个由金属密肋支撑起的仿庑殿玻璃屋面之下，颇有"反宇向天"的意象。空间围绕千年国宝《开成石经》展开，两侧设置连接上下展厅的坡道；大面积玻璃幕墙增加了界面的通透性，光引导着观者行进。整座建筑内部结构清晰，通用性强，可以满足不同的展览需求，同时设置独立的教育、培训、临展等功能，保证在闭馆的情况下仍能对外使用。建筑外部预留城市广场，设置浅水景及城市舞台，适用于举办大型活动。屋顶绿化的设置，为公众提供了视野极佳的观景场地和交流空间。

碑林博物馆是项秉仁继大唐不夜城后，又一次对传统母题的当代演绎。如果说大唐不夜城是传统的"皮"现代的"里"，那碑林博物馆似乎在某种程度上恰恰相反，因此我们可以读到很多类似却更"戏谑"的处理手法，在对城市语境的解读、传统建筑的隐喻和建筑基本元素的表达上无一例外。但无论哪种，它们都属于城市，根植于城市，生长于城市，这一内化的生命历程正是项秉仁设计的初衷。

北侧入口效果 Building front view

Beilin Museum has an important position in the historical development of Xi'an. It is the largest museum of stone steles in China, with a great collection of ancient classic stone inscriptions. Its construction can be dated back to the end of the Song Dynasty, as the museum is an expansion based on the original Confucian temple on site. Through continuous renovation, the architecture has slowly turned into a museum that specializes in collecting, researching and displaying stone steles, epitaphs and stone-carvings. However, this traditional landmark has encountered various problems including insufficient exhibition space, poor preservation condition and equipment aging. In 2017, the Shaanxi Provincial Government and the Xi'an Municipal Government proposed to expand the Beilin Museum and to create a new exhibition hall in the northern district, while renovating its surrounding neighborhood at the same time.

Xiang Bingren believes that the exploration of internal culture should be the catalyst to bind the new architecture with the old. The design aims at immersing the new museum into the urban environment in a more modern and friendly manner. As a result, the historical status of Beilin would be continued and a cultural resonance across time would be established.

The design strengthens the sense of axis at Beilin Museum, reflecting the overall pattern of "one-axis and two-wing". This creates a natural transition between historical and modern spaces, as a metaphor of the interleaving and prolongation of time. The design also draws inspiration from the traditional inscription rubbing technique. The adoption of the modern architecture language transforms the image of stone steles into expressions of architectural skin and space. The bluestone gives a powerful and stable sense to the volume, and the deep and shallow imprints on the facades are mimicking the different pressures in the process of inscription rubbing. Meanwhile, the brick wall responds to the clue of history, while the glass catches the bonding between "ancient" and "modern". The bluestone, glass and brick wall coordinate with each other and work as slices of time, which conserve the authenticity of Beilin as cultural space.

The foyer is located on the ground floor, with a courtyard sitting under the hip-roof-style glass roof. Supported by metal ribs, the roof structure refers to the traditional architectural philosophy in China. Space is oriented around the thousand-year-old national treasure *Kaicheng Stone Classics*, with ramps located on two sides to connect the two-floor gallery. Large glass walls

^ 片区鸟瞰图 Bird's-eyes view

∨ 建筑空间体量与表皮生成 Diagram of spatial volume and building skin

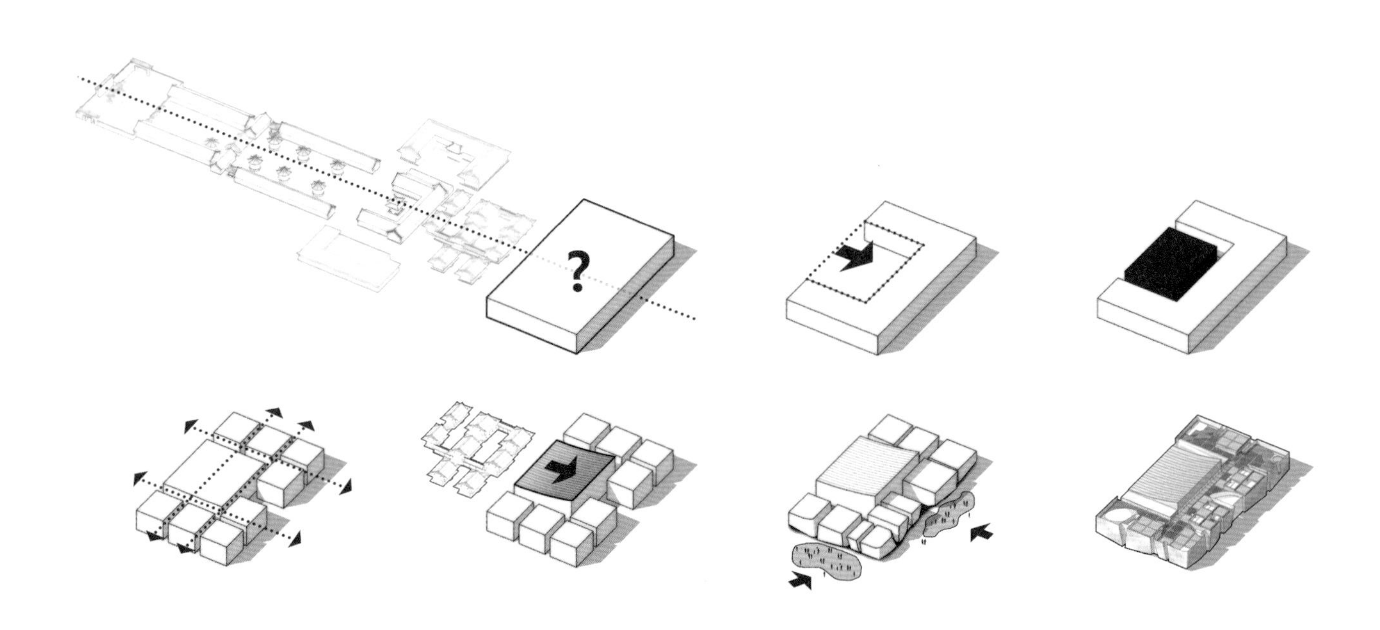

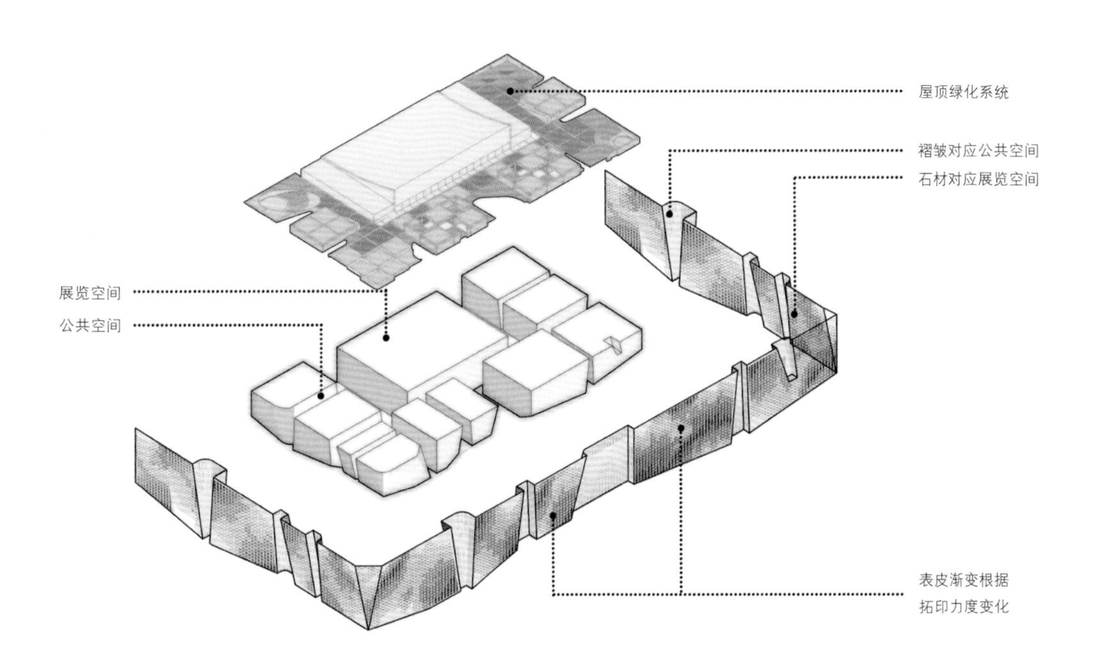

屋顶绿化系统
褶皱对应公共空间
石材对应展览空间
展览空间
公共空间
表皮渐变根据
拓印力度变化

室内大厅效果图 Lobby interior

enhance the translucency of the interface, and the light also helps to guide visitors. The structure of the entire building is clear and versatile, which could meet different exhibition requirements. Besides, separate functional spaces for education, training and temporary exhibition are added, and the design makes sure they could be open to the public even when the museum is closed. The outdoor space is kept as an urban plaza, designed with performance stage to host large events as well as shallow waterscape. The green roof provides the public with an extraordinary viewing platform and space for communication.

Beilin Museum is once again Xiang Bingren's contemporary interpretation of traditional motifs, after his completion of the Great Tang Everbright Town. The Great Tang Everbright Town seems traditional on the outside and modern on the inside. The Beilin Museum is on the contrary, to some extent. Thus, we could identify many similar but more playful design approaches in not only the interpretation of the urban context but also the metaphor of traditional architecture and the expression of basic architectural elements. But anyhow, they all belong to the city, root in the city, and grow in the city. This internalization of the life course is exactly Xiang Bingren's original intention of the design.

# 评论
Review

对于西安碑林这样的有着传统文化底蕴的城市更新课题而言，建筑形式语言的度是很难把握的。设计过程当中项秉仁并不纠结，尝试了多种建筑形式。看淡形式是一般设计师很难达到的境界。然而对项秉仁来说，无论是仿唐风的陕西大剧院文化建筑群，还是完全现代的宁波文化广场，都无不可。风格消隐了，建筑对场地、空间的态度却更为明朗，叙事性在起承转合中慢慢地铺展开来。

整个新馆的核心空间是陈列《开成石经》的展示大厅。石经是博物馆镇馆之宝，完整地保存了迄今所见儒经的最早版本，堪称中华文化的原典。如此重要的一组石碑在高度非常受限的新建场馆条件下，如何能更好地展示出来？这组场馆又如何与博物馆的其他部分联系？项秉仁将展示馆下挖一层，使得展示空间在尺度上开阔起来。一组大台阶将地下公共大堂延伸到《开成石经》展览大厅，并以对称折跑的方式向石经上方的挑空屋宇升腾。屋宇采取的是唐宋大殿的尺度，这种恰到好处的空间体验与《开成石经》无与伦比的地位完美地结合在一起，相得益彰。

新馆的表皮设计也是让人印象深刻的，取法宣纸的石材表皮“拓印”在内部理性的展览空间单元上，形成了凹凸的正负空间。传统书法的形式语言以建筑的虚实关系轻松自然地表现出来，像是写意山水，也像书法长卷。建筑下部的体块倾斜并抬起，印证出项秉仁举重若轻的设计思考。表皮的手法很时髦但富有底蕴，象征着千年来无数拓印碑帖的先人们留下的足迹和思绪。

---

马庆禕

上海秉仁建筑师事务所首席合伙人、总建筑师

This is an urban renewal project targeted at a museum with rich cultural heritages and implications. It can be difficult to grasp the degree of any formal intervention. Yet Xiang did not struggle much in this respect. He experimented with a variety of forms during the design process but didn't get obsessed with any one of them. For him, no particular style is indispensable whether it's the Tang style of Shaanxi Opera House Cultural Complex, or the completely modern style of Ningbo Cultural Plaza. In place of style, perceptions of site and space are becoming clearer, the narrative is gradually formulating throughout the process.

The core space of the new museum is the showroom of the *Kaicheng Stone Classics*, the oldest existing texts of the Confucian classics and undoubtedly the treasure of the museum. How can such a precious group of stone tablets be best displayed in the conditions of new venues with very restricted height? How does this set of venues relate to the rest of the museum? Xiang responds to these questions by digging a layer deeper under the exhibition hall and thus opening up the whole exhibition space. A large staircase connects the underground lobby to the showroom of *Kaicheng Stone Classics*, and then rises to the double-height structure above the *Stone Classics* in a symmetrical way. The structure takes the scale of the grand halls in Tang and Song dynasties, creating a solemn spatial experience that perfectly corresponds to the status of the *Kaicheng Stone Classics*.

The envelope of the new museum is equally impressive. The stone material skin that resembles the texture of Xuan paper has been “lithographically printed” on the internal exhibition spaces, forming concave and convex or positive and negative spaces. The formal language of traditional calligraphy is naturally expressed through architectural reality and virtuality, like a traditional Chinese landscape painting or a calligraphy scroll. The lower part of the building is tilted and raised — a result of Xiang's deep belief in handling complicated matters with ease. The technique of the skin is very fashionable yet rich in heritage, symbolizing the legacies of our ancestors dedicated to rubbing the inscriptions over hundreds of years.

---

Ma Qingyi

Chief Partner/ Executive Chief Architect, DDB Architects Shanghai

# 附录

# Appendix

对谈四则
Four Conversations

项目信息
Project Data

项秉仁履历年表
Xiang Bingren's Chronology

# 对谈四则



# Four Conversations

## 与贝聿铭的一次谈话

1992年是项秉仁在美国学习进修的第3年，贝聿铭亲署来信，通知项秉仁荣获当年的旅美中国学者奖金，并得到一次建筑之旅的资助。当年夏天，项秉仁驱车访问了美国东部的主要城市，考察了大量著名的建筑物、城市设计作品和建筑名校。最后，于1992年8月3日回到纽约，在位于麦迪逊大街600号9楼的贝聿铭先生的办公室与他见面。
下面是这次见面时的谈话记录：

贝聿铭（以下简称贝）：您好！请问您的姓XIANG中文应该怎么写？

项秉仁（以下简称项）：是项羽的项，就是这样（写给贝先生看）。

贝　这次你跑了哪些地方？

项　东部的几个城市，芝加哥、纽约、波士顿、华盛顿特区、费城，还去看了流水别墅。

贝　你还应该去达拉斯，那里有许多好的作品。有路易·康的，菲利普·约翰逊的，我也有几个东西在那里。你在回旧金山的途中可以去一下。改机票并不难，我可以让秘书帮你与航空公司联系，你定的是哪个航空公司的？

项　美国航空公司。

贝　那就好办。我们与那家航空公司有联系。我可以请秘书办。

项　谢谢贝先生！今晚我就要飞回旧金山了，我下次一定去。
（一年后，项秉仁去了达拉斯并参观了贝先生设计的达拉斯市政厅和麦耶生交响中心）

项　我这次来，一则是很想见见贝先生，二则也是来向贝先生表示感谢。

贝　这件事是这样开始的：前些年我了解到一些从国内来的访问学者，由于经费不够，来美国后就只是在一个学校里，半年一年后就回去了，没有能够到美国其他城市看看，这很可惜……

项　是的。建筑一定要亲身体验。任何间接的东西都无法真正传递建筑的体验。

贝　是啊！所以我就想帮助他们。正好我当时得到了一笔奖金（普利兹克奖），我就把这钱作为基金了。其实，所给的钱不多。你知道，建筑师不是富翁，钱只能是小意思。

项　这主要是一个鼓励。

贝　对，算是鼓励吧！所以，用了这笔钱，就是要回国去，不然这就没有意义了。现在许多中国学者来了美国就不回去了。从你的条件看是很不错的。你的夫人还在中国，所以我决定给你这个奖金。

项　我是准备回去的。像我这样的年龄和经历，在中国会发挥较大的作用。不过，贝先生，我想请教你一个问题，许多中国年轻的建筑学生，包括我的一些学生现在都来到了美国，我不知道他们在美国是否能得到发展？不知贝先生怎么看？

贝聿铭，美籍华人建筑师，美国艺术与科学院院士，中国工程院外籍院士。
本文曾收录于《建筑百家回忆录》，中国建筑工业出版社，2000年。录入本书时略有修改。

贝　我看是很难。美国社会是很难打入的。你想想，从中国来美国四五年时间是不够打入美国社会的。美国的文化也不是一朝一夕可以融入的。想想现在有那么多美国建筑师，美国业主为什么一定要用你中国人呢？我来美国达50年了，情况不同。留在美国发展，多数人只能做些小事，做不了大设计。所以，我总是劝他们回中国去。

项　是的。现在中国正在发展中，有许多事情可以做。记得多年前，我在北京参观了贝先生的香山饭店。这在当时确实让我领略了什么是现代建筑。贝先生是不是还有兴趣去中国做些设计？

贝　香山饭店不用提了，弄得不成样子。我都不想再去看了。你想，那些领导干部带着一家子人在那里住，旅馆哪能搞得好经营管理？

香山饭店只是想告诉大家，中国建筑的现代化要找中国自己的办法。中国的情况不同，建高层建筑不一定是唯一的办法。当然高层还是会建，但中国还是要寻找自己的办法。（比起内地，）香港就不同，爱怎么盖就怎么盖。

我不是不想回中国去做设计，只是年纪大了，力不从心。如果有工程在中国，就得经常来来去去。一年跑一两次我还受得了。多了就不行。我手下没有助手，美国人又不了解中国的情况，要有工程就得自己亲自跑，可是又跑不动，所以没法做。

项　贝先生，我这次来东部，看到大城市已经盖了这么多的建筑，是不是美国的建筑还会有高潮？美国建筑的前景如何？

贝　我想，过三五年后情况会有好转，因为旧的总归需要拆除，新的要替代旧的。

项　贝先生，我在波士顿参观了您设计的肯尼迪图书馆，觉得真好，建筑的确成了艺术品。

贝　还是没有做好，因为钱不够了。

项　贝先生，您对自己哪一个作品最满意？

贝　近几年做的东西比较满意。觉得真正是做到了自己想表达的东西。所以我现在还想多做些作品。

项　贝先生现在有哪些工程？

贝　对面的一个旅馆正在建。还有一个工程在日本，一个在西班牙。

项　贝先生有没有作品在旧金山？听说有几个作品。

贝　没有。我现在和公司脱离了，所以常常会搞混。我想在适当的时候要讲讲清楚。

项　是的，有的作品看上去确实不像是您的手笔。贝先生，我还想听听您对上海浦东开发的看法。

贝　上次朱镕基来美国，我出席了。朱镕基先生是个有学问、思路很清楚的人。这样的干部多了，中国的事情会办得好一些。不过，中国的建筑要发展得好，至少还得有二三十年。目前像浦东还是首先得搞好规划，然后是基础设施建设，建筑还是其次。

项　谢谢贝先生的时间。希望贝先生今后还是能去中国看看。

# 与王方戟的一次对谈

对项秉仁教授的访谈发生在一家普通而热闹的咖啡馆内。最初的话题由项教授的一幅住宅装修草图引发。那是一套寻常的四房两厅，其设计围绕的主题是目光的收放与漂移，似乎有意拒绝了铺地、吊顶、护墙等基本室内词汇的介入，将人带入了现代的空间关系之中，也将我们的话题带到了现代主义建筑这个论题上。

王方戟（以下简称王）：在最新的建筑理论研究中我们可以注意到一个有趣的现象，许多建筑师及建筑评论家将目光集中到现代主义建筑刚刚起步的那个时期。更有人将目前中国建筑所面临的问题与德国现代主义建筑雏形时期的问题相提并论。您认为中国的现代建筑与现代主义建筑的关系在哪里?

项秉仁（以下简称项）：我觉得中国建筑的现状是中国特殊的建筑教育发展的结果。50年代的建筑系学生没有条件系统地理解现代主义思想；60年代的学生全面接受了现代主义思想，但却很少有实践的机会；70年代末及以后的学生接受的是后现代多元化思想。所以现代主义建筑理论其实并没有在中国被系统地实践过。这是中国当代建筑面貌产生的根本原因之一。

王　您的话可否理解为中国设计界的混乱状况正是这种不稳定的思想输入的结果?

项　对，脱离对西方现代主义及以后的历史过程的理解，简单狭义地吸收了后现代思想，对中国建筑其实是非常有害的。

王　那您的设计方法是如何形成的?

项　我是60年代的大学生，所以骨子里使用的还是现代主义的逻辑方法。但我发现，现代主义的设计原则常常可以松动，从而给设计添加巨大的自由度。所以如果需要对我设计方法定义的话，我觉得“新现代”还是比较贴切的。

王　您觉得在中国当前的建筑思想状态下，“现代”或“新现代”究竟意味着什么?

项　现代主义建筑思想留给我们的决不是某种特定的形式，而是建筑所特有的形式美如何由逻辑推理来生成。我想举我们最近设计的北京首都博物馆方案作个例子。在描述它的形式时，我几乎无法将其形式与形式的目的剥离。建筑实体前超尺度的柱廊是它最强的特征。它像一个屋顶早已坍塌的“遗迹”，期待着各种可能的功能，并呼应着博物馆的主题。“遗迹”下抬高的平台就像中国式的舞台，造成了建筑与前面长安街的对话关系。平台将人流从北面引到东南面，把光线带到了建筑的入口前。形式与形式的目的之间不断的相互推理构成了现代的设计逻辑。

王　不同的项目有不同的逻辑，作为建筑师如何才能发掘这种逻辑呢?

项　我在东南大学读博士研究生的时候，我的导师童寯先生要求我每周向他汇报两次读书体会，并就不懂的问题向他请教。问题的答案如果是可以在书上查到的就会被批评。这使我深深体会到，提出问题常常比解决问题更关键，更有挑战性。我认为现代设计方法其实也是提示我们寻找项目任务书背后的问题，对这些问题的解答可以归纳成最终逻辑。

王　从您的上海水清木华住宅区设计方案中可以看出住宅按正南北排列的条理，但它是被如何融合在整体逻辑中的呢?

项　现在有些住宅区总体设计跟着感觉走，随意搬弄建筑。从1:1000的图上来看十分乖巧。但它给城市及社区带来的实际效果非常令人怀疑。为了实证住宅区设计所应该具有的现代逻辑性，我完成了这个设计。该社区所处的地块中7层以上的住户可以享受到南面浦东中央公园的景观。所以建筑严格按正南北向排开，并布置出三种空间感截然不同的住区：区前密集的4层，在浦东新区制造出浦西的里弄邻里感，让人体验怀恋与憧憬相互交织的空间；小区东、北外围排开18层的高层，提供了开敞、可以俯瞰住区内景观、并遥望中央公园的居住空间；第三种是围绕社区内部景观的大面积住宅。最关键的是，我们发现当这三种相异的东西各自独立，并理性地叠加在一起之后，区内的空间及体量关系依然很美。

王方戟，同济大学建筑系教授，上海博风建筑设计咨询有限公司主持建筑师。
本文原名《观察与思考——访项秉仁建筑师》，最初发表于《时代建筑》2001（01）：42-45。

王　现在流行一讨论住宅设计就大谈“市场”“卖点”，似乎建筑师的专业技巧已经过时。您的杭州“富春山居”中的别墅设计似乎脱离了市场与建筑师暂时达成的默契。在这个设计中您是如何考虑市场问题呢?

项　在杭州，有人推出了古典式样的美国郊区别墅的完满复制品，卖得不错。一时间这种建筑形式成了衡量市场前景的标尺。这个现象，并不是说西方古典比其它建筑式样好，而恰恰说明我们的建筑师远远没有发挥出让市场惊叹的能量。基于这种考虑，我联想到杭州周围的江南民居。它们虽然经历了许多磨难，今天看起来依然非常漂亮。我的设计借鉴江南民居的形式，并用现代的构造方式加以表现。它与市场主流上的别墅没有许多共通点，但我相信市场没有理由拒绝让人陶醉的美。

王　我发现虽然您一直在谈现代，但作品中却不断流露出与传统的关联。您最近设计的上海复兴公园大门中表现出来的非现代形式与现代形式感的结合让人很感兴趣。在这个设计上您是如何理解传统问题的?

项　传统不应该被简单地理解为“原料库”，而应被看成与现代环境相平等的元素。其意义需要发掘、解释和不断的重新理解。复兴公园是一个浪漫而充满记忆的地方，它最初是法国人的囤兵场，后来被开辟为租界里的公园。它的老大门很有特点，是我童年记忆的一部分，相信也应该是这座城市记忆的一部分。抱着一线希望，我到上海城市档案馆查询历史记录。让人十分激动的是，我竟然找到了老的复兴公园大门的建造图。于是我突发奇想，原封不动地将它搬回来，并想用全新的现代体块将它夹起来。不过最终两旁的体块选择了折衷形式。

王　它像是树的自然生命衰竭了之后，靠种子又生长出同样的形态，演绎了生命的轮回。它既不仿古，又不是真的古董。它介于两者之间的模糊感非常有趣。我觉得这种模糊感使项目的基础条件获得了拓展。那么，您是如何理解建筑设计的前提条件呢?

项　建筑师有许多机会观察这个世界的不同侧面，但我们不能只贪经历不图理解。我觉得对于获悉的信息应该有个抽象的过程。抽象帮助我们拂去现象的尘土，显露深藏的基石。这个过程就应该像一本杂志名称的叫法：《观察与思考》。

# 与薛求理的一次谈话

薛求理（以下简称薛）：我自认为比其他人对项老师的认识要深一点，一直很关注您。

项秉仁（以下简称项）：您不光关注我一人，您对中国大陆的建筑师们都时刻挂心。

薛　项老师的水彩画画得特别好。

项　但我觉得我画得不怎么好（笑）。现在比较冒尖的建筑师，比如王澍、张永和，年轻的时候都画得不错，画画对于提升一个人的建筑修养，还是很有益处的。我总觉得，一个建筑师什么都好是不大可能的，但为了设计建筑需要提高的自己的修养，需要去看书、听音乐、画画、逛博物馆。

薛　项老师读博士期间在《建筑学报》发表了五篇文章，在当时的中国是非常有影响力的。

项　那个时候阅读了很多国外书籍，结合博士论文将读书报告润色了一下去发表，也算是读博士阶段的一个内容。那时做这样事情的人比较少。

薛　工作十年后继续深造，从一个实践者转向学术人，项老师研究的不是空洞的理论，无论是符号学还是环境心理学，目的都是为了指导实践。很少听到项老师大谈理论，但在老师的设计中可以看到理论指导的痕迹。项老师讲话一直比较实际，不会讲和建筑不着边际的东西，就讲建筑师的责任，讲怎么把项目做好，对业主负责。

项　因为个人经历不同。我最近因为出书的关系，也在回忆这些事情。譬如像张永和他们这一代去美国是去读书，受到当时美国建筑教育观念的影响，接受的都是比较新的概念。我们去的时候在国内已经读到博士了，到美国就想参与到实践中去。一旦你加入庞大的建筑师队伍，像弗兰克·盖里这样大师级的人物就很少能遇到，遇到的都是职业建筑师。而且当时的美国，经济高度发达，建筑业已然萧条，对于建筑师来说，服务好每个客户至关重要，丢了一个业主如丧考妣。这种感觉是和在国内高校里不一样的，在大学里面尽管当时中国经济不好，但是建筑师很受尊重，有话语权。

薛　我看项老师的项目，项老师对于每一个时代流行的建筑理念还是很敏感的。有的人就是一个套路来来回回用，但项老师蛮注重这个时代大概什么东西比较流行的。

项　我不是一个咿咿呀呀自我膨胀的人，所以这里面也有一个问题，我老是觉得自己做不好，老是觉得别人的东西比我做得好，老是愿意吸收别人的东西，老是给自己一个追逐的目标吧，这是我与生俱来的一个性格吧。（笑）

薛　我觉得项老师虽然在追赶时代，但有些很俗气的东西从不去碰，是有取舍的，到底还是老师自己的品味啊。

项　希望做得现代一点，现代要做得有匠心，也不是那么简单的。

薛　您现在还参加项目吗？

项　需要我的时候，我会参与进来，基本上是做顾问。

薛　现在你们公司投标还多不多啊？

项　也多，可是没以前多了。

薛　项老师回到上海开业之后，成就很大。中国一共一百多个大剧院，其中有三十多个是外国人设计的，而项老师就设计了两个大剧院。

项　现在还在做陕西大剧院的室内设计。

薛求理，香港城市大学建筑学与土木工程学系副教授。

对谈时间：2017年7月9日
对谈地点：上海秉仁建筑师事务所会议室

薛　像这类竞赛是怎么拿下的?

项　国内的大型项目惯例是请国外的设计师，或者请高校、大型设计院。当时我是以公司的名义参加竞赛的，成绩好就中标了，后期和大设计院合作做施工图。以前我还年轻力壮，投标都可以拿下来。

薛　那时您大概六十几了吧。

项　对的。那时投标都冲上去，几个方案都是我做，一般还可以拿下来，现在就困难了。要做好一个建筑，必须要跑现场好多次。现在很多项目不在上海，投标要去开会，要去了解业主介绍项目的想法，要体会现场的环境，从领导和业主的谈话中揣摩出他们的追求，回来以后没日没夜赶方案图，总是要推翻以前的方案，最后到没有时间了才把方案定下来，再努力去搞效果图，把文本做好，这个过程是很累的。方案中标了以后还要深化方案到初步设计，要开会和领导们推敲。图纸完成到施工建设，要到施工单位把施工图弄好，还要到现场去选择材料，看看是否按照设想的细部在做。一个项目从招投标开始，到最后完成达到理想的效果，这个过程花的精力是非常大的，所以我认为建筑师的体力也是很重要的条件，我自己感觉体力不如人家(笑)，所以认为到了一定年龄就往后退了。以前在纽约遇见贝聿铭，他也是这样讲，做是当然想做，但飞机最多坐个三次，多了就吃不消啦。什么年纪该做什么样的事情嘛。

薛　你们同学里面还有在做的吗?

项　几乎没有了，像我这样兼做设计的就更少了。因为毕业的时候"文化大革命"，很多人没有项目可做，改行去做管理。

薛　项老师做了很多大的工程。建筑媒体对您这样的建筑师关注不够多，作为一个旁观者，有时候会觉得媒体有点偏颇不公。

项　杂志肯定要选夺人眼球的东西嘛。建筑学这个学科呢，涵盖的方面太大，什么角度都可以，真的很难用一个固定的标准去评价。客观来说，我没有新一代建筑师的锐气和才气，但是在老一代里面，我还算是走在前面一点的。

# 与张永和的一次对谈

2018年6月，上海暴雨，项秉仁与张永和重逢在一间普通的工作室，回忆起各自人生轨迹的许多重合点。对谈从1978年在南京工学院的相识讲起。

张永和（以下简称张）：我到南京是1978年的春天。你是春天入校的吗？

项秉仁（以下简称项）：我是1978年秋天。

张　噢，我是读本科，老项是读研究生。我们是七七级，春天入学不是正常的。有一个场景到现在我还记得，我们到楼上的一个房间，屋顶是斜的，类似上海有老虎窗的亭子间这样一个空间，看到三位研究生在那里画图。老项你是坐在左手边，在画一个小剧院。

项　那应该是1979年了。

张　当时是带着崇敬的心情去看你们的，学校那时候没有高年级学生。老项是在画水粉，坐在右手边的是黎志涛，在用尺规铺墨线，画夜景。

项　黎志涛后来留在学院当老师了。1978年是“文化大革命”之后第一年研究生招生，当初很多人不敢考，觉得很难考。那时候导师也很少，对导师的要求也很高，像刘光华教授、齐康教授、钟训正教授这样的老师也没有资格一人带一名研究生，要三人联合带三名研究生。所以建筑设计专业就收了我们三个人：我、黎志涛和仲德崑。
当时画那个小剧院，我用的是水粉，那时候在建筑系画水粉表现的人还是比较少的，大多数人是画水彩。黎志涛是用钢笔画，画的是夜景，他是用线条把背景涂黑，把星星留白出来，非常不容易。

张　总之我们就是这么认识的。我们班还有七八级那班有一个共识，老项代表了南工（南京工学院）精神的传承。

项　我认识张老师呢，是特意去本科生的教室看他们。当时知道张永和进南工之前在北京学过油画，就对他印象很深。我对会画油画的人很敬佩，因为觉得很难画。

张　那你当时也没告诉我（笑）。

项　你离开（学校）特别早。三年之后就去美国了，当时也是一个新闻。

张　因为这事，还学会了个字，“肄业”的“肄”字（笑）。你是什么时候去的旧金山啊？

项　我是1989年。

张　我是1990年去伯克利教了两年书。我工作过的事务所离你工作的事务所好像只隔一两个街口。你们事务所的玻璃特别黑，我有这么一个印象。

项　我和张老师在旧金山的时候来往比较多，他自己的作品展览会邀请我去看。张老师做过一些很特别的事情，比如“吃面条”装置设计，这个设计还得了奖，还有关于自行车的设计，当时觉得跟我们在国内的观念就相差很大了。

张　好像有一次我们聊到要做什么样的建筑，老项你说要做中国的SOM，还记得吗？

项　吹牛吧（笑）。

张　不是，不是说要做成SOM的规模，这里面其实有一个态度在的。到现在我做设计的时候还会纠结怎么去平衡建筑的一些基本和不那么基本的问题。能把基本问题，如一个节点、一个细节处理得比一般情况更好，就是我理解的老项说的要做SOM的意思。若

张永和，同济大学教授，非常建筑主持建筑师。

对谈时间：2018年6月20日
对谈地点：非常建筑上海工作室

干年以后，有一天晚上在上海咱们吃完饭，你带我和鲁力佳去看复兴公园大门，看过之后我又想起你说过的那句话。那个时候我刚对建造有意识，对于我们建筑师来说，要真的考虑到很细的程度，就能够控制住质量。

项 在美国的时候蛮有意思的，缪朴、张永和和我经常聚，会把做的东西拿出来看。张老师在美国有那么多成就，跟他许多新的理念是分不开的。当时他跟我讲，他不满足于当大学老师，他要做系主任。当时我都觉得他有点吹牛了（笑），但是他后来做到了。那时候也聊下一步怎么走，我跟他讲，你应该回国去，到了国内肯定能成为中国建筑师新生代的代表，在国际上产生更大的影响。

张 这么回忆让我想到这么个故事。你拿过的那个贝聿铭旅美奖学金，我爸也拿过，他是拿这个奖金的第一个人。然后他就去了波士顿，去了哈佛。我爸跟我说贝先生的合伙人亨利 ·N. 柯博（Henry N. Cobb）在纽约做实践，也在哈佛大学建筑系当系主任。那时候年轻气盛，我就跟我爸讲，我以后也想当系主任。我爸当时的表情就像是我说错话了，我反思可能是我野心太大了，他说不是，你野心太小，你应该想着去做个好建筑师。

项 后来我去了香港，咱们在香港也见过。你看到我做的事情，然后你说了一句，这是我爸爸想让我做的事情。（笑）

1990年代，是中国“实验建筑”如火如荼的年代，1993年建立非常建筑的张永和更是这股大潮的领军人物。与此同时，项秉仁全面走向“职业建筑师”的道路。

张 首先建筑师们想的肯定是不一样的，但有些认识又是大家共有的。老南工在这里起到了一个比较积极的作用，建立了一套对于建筑的基本认识，最终还是要盖房子，想方设法把房子盖得好一点。所以在这个基础上谈分歧，就会发现这类分歧并不是最主要的。“概念”对设计是很重要，但是房子完成之后，概念离开现场，就不重要了。当然，如果概念能够帮你想建筑，那也很好。

项 我感觉建筑教育与建筑师的道路选择有非常紧密的关联。我和张老师都是在东南大学读的书，在我的印象中，包括王澍，大家的美术底子都是很好的。我们这一代人，布扎式的学院教育是根深蒂固的，这种影响不光是说你随便勾勾画画就能画出一个很好的东西来，而且是反映了一种美术修养和基本功。这种基本功对人很重要，但是不能成为限制建筑概念发展的一种束缚。我们几个当初在学校里都是基本功很好的学生，但是后来走的道路不一样，比较重要的原因还是教育的不同。我是在东南大学完成了从本科到博士的教育，张老师三年级以后到了美国。当时中国和美国的建筑学教育有很大的差距，美国的整体比较先进，他们经过了我们还没有经历的过程。

我们国内基本上还是职业化的基础教育，包括杨廷宝、童寯这样的老教授，他们的专业实践基本上也是开个事务所，做一些公共或商业性的建筑。虽然也有一些探索，但还是作为一名职业建筑师去满足业主的需求，完成一个建筑设计的合同。所以对于我们，建筑的基本问题就是功能完善、有形式感、结构合理等方面。但是我到了美国以后，发现美国的年轻建筑师会很不一样。为什么呢？还是跟他受到的建筑教育有关系。张老师到了美国以后，很快接受了这样的教育，还有了自己的创意。而我到了美国，就直接到事务所投身实践了。职业建筑师有自己的责任，可能个人的东西就考虑得比较少一点。

其实我很欣赏张老师他们这一代“实验建筑师”做的设计，也很愿意去尝试，但一是机会比较少，二是没有受过这样的教育引导，完全靠自发也是不容易的。所以我后来就没走上这条道路。张老师也拉过我几次，有一次是建川博物馆，后来我做的方案，可惜没建。

张 其实没有哪个时期比另外一个时期好这回事。我对古典主义很感兴趣，像我听巴赫挺多的，但你很难跨越300多年去想象巴赫当时的生活，他可能是每天朝九晚五去教堂弹琴？但古典音乐就是有超越时代的魅力。我爸爸也是南工的，家里还有他两幅当时的作业。他画的内容我并不感兴趣，但是他画的方法很有意思。一个古典平面，一定是铺地都画出来，不会是空白的，铺地砖全画得清清楚楚；还有光线，能把朝夕、方向都表示出来。我也画过一些古典渲染，后来就是拿个速写本草草画画了，这里面确实有时间成本的问题。

我做建筑师的基因里有古典主义，是在南工学的。为什么要一遍一遍渲染，是有制作的功夫在里头，其实对思维的节奏影响很大，只看其实远远不够。要能有感受，一定要画。

关于现代城市和当代建筑的看法。

张　1992年，我拿了一个旅行奖金去欧洲，定的旅行计划就是去看那些书、杂志、幻灯片里见过的现代建筑。结果一到欧洲，发现对人的生活有决定意义的其实是城市环境。欧洲的城市都特别宜居，它有步行的尺度关系，而现代建筑完全可以插入到古老的城市里去。所以我从那会儿就把现代城市在脑子里给否定了。今天如果我要做城市设计，很可能设计一个古典的、传统的城市，会用周边式的街区这种概念。帕特里克·舒马赫让我一定要去看巴西利亚，他说那里验证了现代城市也可以是宜居的。我也没去看过，虽说我听他讲还挺有意思，但还是没有扭转我的城市观念。所以对我来说，城市需要向后看。建筑我还是把它当作城市的一部分，不是一个孤立的物体。

项　我最近有一些比较悲观的想法。以前作为建筑师，作为一个关心城市的人，希望能看到一个美丽的城市，但是跑过了世界很多地方，会发现这座城市名气很大，也能看到很多让人失望的地方，美丽的城市是可望而不可即的。

我是建筑师，对城市研究也不多，跟做规划的人观点更加南辕北辙。现在国内做的很多城市规划，都是一个套路，大小城市都是一根主线，几个点，这样建出来的城市都是千篇一律的。我还是主张城市要自发生长，而不是人为去塑造一个城市形态。现在的城市设计就是先做模型、做体块，摆一摆看起来不错，等变成真的房子之后就很差。

张　今天规划城市的人完全是抽象的操作，完全没有想象力，一堆数字，在图上摆对了就行。就没有真的想象一下，人在街上走，旁边店家开了门，桌椅摆出来这种生活场景。

项　就算是一个好的城市设计也不等于有一个好的城市。好的城市还是要靠建筑细节和一些城市景观、细部来去支撑，可能我还是因为自己是建筑学出身，所以更注重这些微观的东西。宏观上我觉得城市规划不是一个专业人士就能把握的，城市设计还有政治性和社会性。政治家和城市领导是很大的干预因素。

关于服务业主、服务客户。

张　刚才老项也提到杨廷宝、童寯这一代人都很有服务客户的意识。这其实也包含在巴黎美术学院对于建筑的认识当中，建筑师在接触社会的时候界面是很清楚的，在这个基础上去做社会实践。所谓服务业主，简单点说可能就是要像医生一样，建筑师是有责任的。

但是对我来说，还有一个在这之上的追求，是建筑学。建筑总是有一个使用者，但不见得一定是完全服务于这个人，是不是有可能我们提供的是一种更好的体验？时间感、空间感、光线感之类的。不过这样做肯定会有冲突，建筑学有自己的一个参考系，但是业主没有。所谓建筑学的追求，我从来也不跟业主讲，也不是该不该讲的问题，而是你讲了会让他更困惑。不过慢慢时间长了，和业主之间互相了解了，现在的关系就改善很多。

项　我觉得建筑师的责任就是要给业主寻找一个解决他的问题的最好的答案。这个答案不一定是他想要什么东西，我就给他什么东西。有时候我可以提供更好的答案。这跟简单的服务业主不同，简单的服务业主很消极。所谓最好的解决方案，一是在建筑学方面有追求，二是在解决问题的同时，解决了其他的社会问题、城市问题，把其他人的需求和社会的公平性都考虑在里面。不过参加一些政府的大型公共建筑投标时，也会有思想斗争，领导喜欢象征意义，喜欢一些愿景描述。我觉得好的东西，拿出去领导不喜欢，这个时候我会犹豫，有时候会转到这个方面考虑多一些。因为觉得自己花了这么多努力，不中标等于一场空，中标还可以争取再做得更加符合我的想法一点。

# 项目信息



## Project Data

**雨山湖公园小品建筑**

| | |
|---|---|
| 类型 | 景观小品 |
| 位置 | 安徽省马鞍山市 |
| 时间 | 1976 / 1976 |
| 建筑面积 | 400m$^2$ |
| 业主 | 马鞍山市政局 |
| 设计单位 | 马鞍山市建筑设计院 |

**Landscape Architecture of Yushanhu Park**

| | |
|---|---|
| Function | Landscape Piece |
| Location | Ma'anshan, Anhui Province |
| Design / Completion | 1976 / 1976 |
| Floor Area | 400m$^2$ |
| Client | Ma'anshan Municipal Government |
| Design Unit | Ma'anshan Architectural Design Institute |

**马鞍山富园贸易市场**

| | |
|---|---|
| 类型 | 专业市场 |
| 位置 | 安徽省马鞍山市 |
| 时间 | 1984 / 1985 |
| 建筑面积 | 3 000m$^2$ |
| 业主 | 马鞍山市政府 |

**Fuyuan Market, Ma'anshan**

| | |
|---|---|
| Function | Market |
| Location | Ma'anshan, Anhui Province |
| Design / Completion | 1984 / 1985 |
| Floor Area | 3,000m$^2$ |
| Client | Ma'anshan Municipal Government |

**昆山鹿苑市场**

| | |
|---|---|
| 类型 | 专业市场 |
| 位置 | 江苏省昆山市 |
| 时间 | 1986 / 1986 |
| 建筑面积 | 5 000m$^2$ |
| 业主 | 昆山市工商局 |

**Luyuan Market, Kunshan**

| | |
|---|---|
| Function | Market |
| Location | Kunshan, Jiangsu Province |
| Design / Completion | 1986 / 1986 |
| Floor Area | 5,000m$^2$ |
| Client | Kunshan Industry and Commerce Bureau |

**胡庆余堂药业旅游区规划**

| | |
|---|---|
| 类型 | 旅游区规划 |
| 位置 | 浙江省杭州市 |
| 时间 | 1987 / |
| 建筑面积 | 30 000m$^2$ |
| 业主 | 杭州市规划局 |
| 设计参与 | 俞霖、吴长福、包小枫 |
| 设计单位 | 同济大学建筑城规学院 |

**Hu Qing Yu Tang Pharmaceutical Tourist Area**

| | |
|---|---|
| Function | Tourist Area Planning |
| Location | Hangzhou, Zhejiang Province |
| Design / Completion | 1987 / |
| Floor Area | 30,000m$^2$ |
| Client | Hangzhou Urban Planning Bureau |
| Design Team | Yu Lin, Wu Changfu, Bao Xiaofeng |
| Design Unit | College of Architecture and Urban Planning, Tongji University |

**上海复兴公园园门重建**

| | |
|---|---|
| 类型 | 公园小品 |
| 位置 | 上海市 |
| 时间 | 2000 / 2000 |
| 业主 | 上海市园林管理局 |
| 设计参与 | 岳奎、谢印新 |

**Fuxing Park Gate Renovation**

| | |
|---|---|
| Function | Landscape Piece |
| Location | Shanghai |
| Design / Completion | 2000 / 2000 |
| Client | Shanghai Landscape Bureau |
| Design Team | Yue Kui, Xie Yinxin |

**雨花台南大门建筑**

类型　公园小品
位置　江苏省南京市
时间　2001 / 2001
业主　南京雨花台烈士陵园管理局
设计参与　周关良、周凌、倪频频、谢印新

**South Entrance of Yuhuatai Memorial Park**

Function　Landscape Piece
Location　Nanjing, Jiangsu Province
Design/ Completion　2001 / 2001
Client　Nanjing Yuhuatai Martyrs Cemetery Authority
Design Team　Zhou Guanliang, Zhou Ling, Ni Pinpin, Xie Yinxin

**杭州中山中路历史街区保护与更新**

类型　城市设计
位置　浙江省杭州市
时间　2006 / 2007
建筑面积　121 000m$^2$
业主　杭州市规划局
设计参与　吴欣、祁涛、梁贵堡、马庆禕 等

**Protection and Renovation for the Historic Neighborhood of Mid. Zhongshan Rd., Hangzhou**

Function　Urban Design
Location　Hangzhou, Zhejiang Province
Design/ Completion　2006 / 2007
Site Area　121,000m$^2$
Client　Hangzhou Urban Planning Bureau
Design Team　Wu Xin, Qi Tao, Liang Guibao, Ma Qingyi, et al.

**杭州元福巷历史街区保护更新**

类型　城市更新
位置　浙江省杭州市
时间　2006 /
建筑面积　45 940m$^2$
业主　杭州市居住区发展中心
　　　杭州银嘉房地产开发有限公司
设计参与　吴欣、张俊、李娜、肖志抡 等

**Conservation and Renovation for the Yuanfuxiang Historic Block, Hangzhou**

Function　Urban Renovation
Location　Hangzhou, Zhejiang Province
Design / Completion　2006 /
Floor Area　45,940m$^2$
Client　Hangzhou Residential Development Center
　　　Hangzhou Yinjia Real Estate Development Co., Ltd.
Design Team　Wu Xin, Zhang Jun, Li Na, Xiao Zhilun, et al.

**杭州富春山居别墅区**

类型　居住建筑
位置　浙江省杭州市
时间　2000 / 2004
建筑面积　300 000m$^2$
业主　浙江金都房地产开发有限公司
设计参与　秦戈今、缪琦、殷星远、吴刚、练瑞桐、倪频频、李娜 等
合作设计　浙江大学建筑设计研究院

**Fuchun Mountain Villa, Hangzhou**

Function　Residential Building
Location　Hangzhou, Zhejiang Province
Design/ Completion　2000 / 2004
Floor Area　300,000m$^2$
Client　Zhejiang Jindu Real Estate Development Co., Ltd.
Design Team　Qin Gejin, Miao Qi, Yin Xingyuan, Wu Gang, Lian Ruitong, Ni Pinpin, Li Na, et al.
Collaborator　The Architectural Design & Research Institute of Zhejiang University Co., Ltd.

**广州大一山庄**

类型　居住建筑
位置　广东省广州市
时间　2006 / 2009
建筑面积　766m$^2$
业主　广州高雅房地产开发有限公司
设计参与　徐震、肖志抡 等
合作设计　五合国际建筑设计集团

**Dayi Mountain Villa, Guangzhou**

Function　Residential Building
Location　Guangzhou, Guangdong Province
Design / Completion　2006 / 2009
Floor Area　766m$^2$
Client　Guangzhou Gaoya Real Estate Development Co., Ltd.
Design Team　Xu Zhen, Xiao Zhilun, et al.
Collaborator　Wuhe International Architectural Design Corporation

**千岛湖润和建国度假酒店**

| | |
|---|---|
| 类型 | 度假酒店 |
| 位置 | 浙江省淳安县 |
| 时间 | 2004 / 2012 |
| 建筑面积 | 63 453m$^2$ |
| 业主 | 浙江淳安千岛实业投资有限公司 |
| 设计参与 | 彭黎娟、蔡沪军 等 |
| 合作设计 | 浙江省建筑设计研究院（施工图设计） |
| 项目摄影 | 曹呈祥 |

**Qiandaohu Runhe Jianguo Hotel**

| | |
|---|---|
| Function | Resort Hotel |
| Location | Chun'an County, Zhejiang Province |
| Design / Completion | 2004 / 2012 |
| Floor Area | 63,453m$^2$ |
| Client | Zhejiang Chun'an Qiandao Industrial Investment Co., Ltd. |
| Design Team | Peng Lijuan, Cai Hujun, et al. |
| Collaborator | Zhejiang Institute of Architectural Design (Construction drawing design) |
| Photographer | Cao Chengxiang |

**招商 · 万科佘山珑原别墅**

| | |
|---|---|
| 类型 | 居住建筑 |
| 位置 | 上海市松江区 |
| 时间 | 2010 / 2012 |
| 建筑面积 | 176 214m$^2$ |
| 业主 | 上海静园房地产开发有限公司 |
| 设计参与 | 吴欣、郑滢、肖俊瑰、颜莺、王瑜珞、王欢、李晓军、朱莹 等 |
| 合作设计 | 上海三益建筑设计有限公司 |
| 项目摄影 | 吕恒中、曹呈祥 |

**Sheshan Longyuan Villa**

| | |
|---|---|
| Function | Residential Building |
| Location | Songjiang District, Shanghai |
| Design / Completion | 2010 / 2012 |
| Floor Area | 176,214m$^2$ |
| Client | Shanghai Jingyuan Real Estate Development Co., Ltd. |
| Design Team | Wu Xin, Zheng Ying, Xiao Jungui, Yan Ying, Wang Yuluo, Wang Huan, Li Xiaojun, Zhu Ying, et al. |
| Collaborator | Sanyi Architectural Design Co., Ltd. |
| Photographer | Lyu Hengzhong, Cao Chengxiang |

**旺山六境**

| | |
|---|---|
| 类型 | 居住建筑 |
| 位置 | 江苏省苏州市旺山景区 |
| 时间 | 2011 / |
| 建筑面积 | 1 595m$^2$ |
| 业主 | 苏州吴中集团 |
| 设计参与 | 马庆禕 |

**The Six Villas, Wangshan**

| | |
|---|---|
| Function | Residential Building |
| Location | Wangshan Scenic Area, Suzhou, Jiangsu Province |
| Design / Completion | 2011 / |
| Floor Area | 1,595m$^2$ |
| Clients | Suzhou Wuzhong Group Co., Ltd. |
| Design Team | Ma Qingyi |

**江苏电信业务综合楼**

| | |
|---|---|
| 类型 | 办公建筑 |
| 位置 | 江苏省南京市 |
| 时间 | 1998 / 2002 |
| 建筑面积 | 85 000m$^2$ |
| 业主 | 江苏省电信有限公司 |
| 合作设计 | 江苏省建筑设计研究院有限公司 |
| 项目摄影 | 郭新新 |

**Jiangsu Telecom Multi-functional Building**

| | |
|---|---|
| Function | Office Building |
| Location | Nanjing, Jiangsu Province |
| Design / Completion | 1998 / 2002 |
| Floor Area | 85,000m$^2$ |
| Client | Jiangsu Telecom Company Co., Ltd. |
| Design Unit | Jiangsu Provincial Architectural D&R Institute Co., Ltd. |
| Photographer | Guo Xinxin |

**南京电视电话综合楼**

| | |
|---|---|
| 类型 | 办公建筑 |
| 位置 | 江苏省南京市 |
| 时间 | 1998 / 2002 |
| 建筑面积 | 40 000m$^2$ |
| 业主 | 南京市电信局 |
| 设计参与 | 秦戈今 |
| 合作设计 | 南京市建筑设计研究院 |
| 项目摄影 | 郭新新 |

**Nanjing Telephone and Televison Service Center Building**

| | |
|---|---|
| Function | Office Building |
| Location | Nanjing, Jiangsu Province |
| Design / Completion | 1998 / 2002 |
| Floor Area | 40,000m$^2$ |
| Client | Nanjing Telecommunication Bureau |
| Design Team | Qin Gejin |
| Collaborator | Nanjing Institute of Architectural Design and Research |
| Photographer | Guo Xinxin |

**深圳中央商务大厦**

| | |
|---|---|
| 类型 | 办公建筑 |
| 位置 | 广东省深圳市 |
| 时间 | 2001 / 2003 |
| 建筑面积 | 45 000m$^2$ |
| 业主 | 深圳市规划局 |
| 设计参与 | 秦戈今、倪频频 |
| 合作设计 | 机械工业第一设计研究院深圳分院 |
| 项目摄影 | 曹呈祥 |

**Shenzhen Central Business Tower**

| | |
|---|---|
| Function | Office Building |
| Location | Shenzhen, Guangdong Province |
| Design / Completion | 2001 / 2003 |
| Floor Area | 45,000m$^2$ |
| Client | Urban Planning, Land & Resources Commission of Shenzhen Municipality |
| Design Team | Qin Gejin, Ni Pinpin |
| Collaborator | First Design & Research Institute, MI China Shenzhen Branch |
| Photographer | Cao Chengxiang |

**江苏移动通信枢纽工程**

| | |
|---|---|
| 类型 | 办公产业建筑 |
| 位置 | 江苏省南京市 |
| 时间 | 2002 / 2006 |
| 建筑面积 | 48 000m$^2$ |
| 业主 | 江苏移动通信有限责任公司 |
| 设计参与 | 董立军、韩冰、郑泳 |
| 合作设计 | 江苏省建筑设计研究院有限公司 |
| 项目摄影 | 郭新新 |

**Jiangsu Mobile Communication Complex**

| | |
|---|---|
| Function | Office & Industrial Building |
| Location | Nanjing, Jiangsu Province |
| Design / Completion | 2002 / 2006 |
| Floor Area | 48,000m$^2$ |
| Client | Jiangsu Mobile Communication Co., Ltd |
| Design Team | Dong Lijun, Han Bing, Zheng Yong |
| Collaborator | Jiangsu Provincial Architectural D&R Institute Co., Ltd |
| Photographer | Guo Xinxin |

**深圳天利中央商务大厦**

| | |
|---|---|
| 类型 | 办公建筑 |
| 位置 | 广东省深圳市 |
| 时间 | 2003 / 2005 |
| 建筑面积 | 247 449m$^2$ |
| 业主 | 深圳市天利置地有限公司 |
| 设计参与 | 董立军、黄华 |
| 合作设计 | 机械工业第一设计研究院深圳分院 |
| 项目摄影 | 曹呈祥 |

**Tianli Central Business Tower, Shenzhen**

| | |
|---|---|
| Function | Office Building |
| Location | Shenzhen, Guangdong Province |
| Design / Completion | 2003 / 2005 |
| Floor Area | 247,449m$^2$ |
| Client | Shenzhen Tianli Land Co., Ltd. |
| Design Team | Dong Lijun, Huang Hua |
| Collaborator | First Design & Research Institute, MI China Shenzhen Branch |
| Photographer | Cao Chengxiang |

**上海建汇大厦改造**

| | |
|---|---|
| 类型 | 城市更新 |
| 位置 | 上海市 |
| 时间 | 2006 / 2011 |
| 建筑面积 | 39 846m$^2$ |
| 业主 | 上海市徐汇区商业建设总公司 |
| 设计参与 | 秦戈今、王欢 |
| 合作设计 | 上海长福工程结构设计事务所（结构顾问）<br>上海核工程研究设计院建筑设计院（施工图设计）<br>沈阳远大铝业工程有限公司（幕墙设计） |
| 项目摄影 | 曹呈祥、黄紫璇 |

**Jian Hui Tower Renovation, Shanghai**

| | |
|---|---|
| Function | Urban Renovation |
| Location | Shanghai |
| Design / Completion | 2006 / 2011 |
| Floor Area | 39,846m$^2$ |
| Client | Shanghai Xuhui District Commercial Construction Corporation |
| Design Team | Qin Gejin, Wang Huan |
| Collaborators | Shanghai Changfu Engineering Structure Design Firm (Structural advisory)<br>Shanghai Institute of Nuclear Engineering Architectural Design Institute (Construction drawing design)<br>Shenyang Yuanda Aluminum Engineering Co., Ltd. (Curtain wall design) |
| Photographer | Cao Chengxiang, Huang Zixuan |

### 宁波市东部新城行政中心

类型　办公建筑
位置　浙江省宁波市
时间　2006 / 2012
建筑面积　261 365m$^2$
业主　宁波市东部行政区建设指挥部
　　宁波新城服务投资有限公司
设计参与　吴欣、秦戈今、张峻、吴波、王欢、颜莺 等
合作设计　宁波城建设计研究院（施工图设计）
项目摄影　吕恒中

### 宁波城乡建委大楼

类型　办公建筑
位置　浙江省宁波市
时间　2008 / 2011
建筑面积　77 460m$^2$
业主　宁波城建投资控股有限公司
设计参与　史晨鸣、王瑜珞、滕露莹
合作设计　宁波市城建设计研究院有限公司
　　（施工图设计）
项目摄影　曹呈祥

### 宁波电力大楼

类型　办公产业建筑
位置　浙江省宁波市
时间　2008 / 2012
建筑面积　60 700m$^2$
业主　宁波市电力局
设计参与　史晨鸣、马庆禕、王瑜珞
合作设计　浙江高专建筑设计研究院有限公司
　　（施工图设计）

### 招商局上海中心

类型　办公建筑
位置　上海市
时间　2012 / 2016
建筑面积　43 031m$^2$
业主　招商局蛇口工业区控股股份
　　有限公司上海公司
设计参与　马庆禕、王瑜珞、崔美兰、黄立妙
合作设计　上海建筑设计研究院有限公司
　　（施工图设计）
　　上海栖地景观规划设计有限公司
　　（景观设计）
项目摄影　存在建筑

### 合肥大剧院

类型　文化建筑
位置　安徽省合肥市
时间　2003 / 2009
建筑面积　58 776m$^2$

### Ningbo New Municipal Centre

Function　Office Building
Location　Ningbo, Zhejiang Province
Design / Completion　2006 / 2012
Floor Area　261,365m$^2$
Clients　Ningbo New Municipal Centre Construction Headquarters
　　Ningbo New Town Service & Investment Corporation
Design Team　Wu Xin, Qin Gejin, Zhang Jun, Wu Bo, Wang Huan, Yan Ying, et
Collaborator　Ningbo Architecture Design Institute (Construction drawing des
Photographer　Lyu Hengzhong

### Ningbo Housing & Urban-Rural Committee Building

Function　Office Building
Location　Ningbo, Zhejiang Province
Design / Completion　2008 / 2011
Floor Area　77,460m$^2$
Client　Ningbo Urban Construction Investment Holding Co., Ltd.
Design Team　Shi Chenming, Wang Yuluo, Teng Luying
Collaborator　Ningbo Urban Construction Design Institute
　　(Construction drawing design)
Photographer　Cao Chengxiang

### Ningbo Electric Power Authority Building

Function　Office & Industrial Building
Location　Ningbo, Zhejiang Province
Design / Completion　2008 / 2012
Floor Area　60,700m$^2$
Client　Ningbo Electric Power Bureau
Design Team　Shi Chenming, Ma Qingyi, Wang Yuluo
Collaborator　Zhejiang College of Architecture Design and Research
　　Institute Co., Ltd. (Construction drawing design)

### China Merchants Group Shanghai Center

Function　Office Building
Location　Shanghai
Design / Completion　2012 / 2016
Floor Area　43,031m$^2$
Client　CMSK Shanghai Branch
Design Team　Ma Qingyi, Wang Yuluo, Cui Meilan,
　　Huang Limiao
Cooperation　Shanghai Institute of Architectural Design &
　　Research Co., Ltd.(Construction drawing design)
　　Shanghai QIDI Design Group
　　(Landscape design)
Photographer　Arch-Exist

### Hefei Grand Theatre

Function　Cultural Building
Location　Hefei, Anhui Province
Design / Completion　2003 / 2009
Floor Area　58,776m$^2$

| 业主 | 合肥市政务文化新区建设指挥部<br>合肥政务文化新区开发投资有限公司 |
| --- | --- |
| 设计参与 | 董屹、吴欣、张俊、程翌、韩冰、<br>滕露莹、郑滢 等 |
| 合作设计 | 同济大学建筑设计研究院（施工图设计）<br>同济大学建筑声学所（声学顾问）<br>同济大学建筑城规学院 - 视觉与照明研究中心（灯光设计）<br>总装备部工程设计研究总院（舞台设备顾问）<br>北京五合国际建筑设计咨询有限公司（建筑节能设计）<br>深圳新科特种装饰工程公司（室内施工图设计） |
| 项目摄影 | 张嗣烨、合肥大剧院 |

| Clients | Hefei New Municipal and Cultural District Construction Headquarters<br>Hefei New Municipal and Cultural District Development Investment Co., Ltd. |
| --- | --- |
| Design Team | Dong Yi, Wu Xin, Zhang Jun, Cheng Yi, Han Bing, Teng Luying, Zheng Ying, et al. |
| Collaborators | The Architectural Design & Research Institute of Tongji University (Construction drawing design)<br>Tongji University Institute of Building Acoustics (Acoustics consultant)<br>Vision and Lighting Research Center at College of Architecture and Urban Planning,Tongji University (Lighting design)<br>General Armament Department Engineering Design and Research Institute (Stage equipment consultant)<br>Beijing Wuhe International Architectural Design Consulting Co., Ltd. (Building energy efficiency design)<br>Shenzhen Special Decorative Engineering Company (Interior construction drawing design) |
| Photographer | Zhang Siye |

**西安大唐不夜城贞观文化广场**

| 类型 | 文化建筑 |
| --- | --- |
| 位置 | 陕西省西安市 |
| 时间 | 2005 / 2017 |
| 建筑面积 | 101 300m$^2$ |
| 业主 | 西安曲江新区土地储备中心<br>西安曲江文化产业投资（集团）有限公司 |
| 设计参与 | 陈强、程翌、滕露莹、郑滢、王宝卓、<br>王欢、李晓军 等 |
| 合作设计 | 西北建筑设计研究院（施工图设计）<br>马歇尔戴声学有限公司<br>Marshall Day acoustics（声学顾问）<br>总装备部工程设计研究总院（舞台设备顾问） |
| 项目摄影 | 存在建筑 |

**Great Tang Everbright Town Zhenguan Cultural Center, Xi'an**

| Function | Cultural Building |
| --- | --- |
| Location | Xi'an, Shaanxi Province |
| Design / Completion | 2005 / 2017 |
| Floor Area | 101,300m$^2$ |
| Client | Xi'an Qujiang New District Land Reserve Center<br>Xi'an Qujiang Cultural Industry Investment (Group) Co., Ltd. |
| Design Team | Chen Qiang, Cheng Yi, Teng Luying, Zheng Ying, Wang Baozhuo, Wang Huan, Li Xiaojun, et al. |
| Collaborators | Northwest Architectural Design and Research Institute (Construction drawing design)<br>Marshall Day Acoustics Co., Ltd. (Acoustics advisory)<br>General Armament Department Engineering Design and Research Institute (Stage equipment advisory) |
| Photographer | Arch-Exist |

**陕西大剧院**

| 类型 | 文化建筑 |
| --- | --- |
| 位置 | 陕西省西安市 |
| 时间 | 2005 / 2017 |
| 建筑面积 | 52 324m$^2$ |
| 业主 | 西安曲江新区社会事业管理服务中心<br>西安曲江大唐不夜城文化商业（集团）有限公司 |
| 设计参与 | 滕露莹、郑滢、王欢、李晓军、王瑜珞、许江琦、<br>李娜、黄立妙、瞿子岚、李江、高凯、顾金娣 等 |
| 合作设计 | 西北建筑设计研究院（施工图设计）<br>浙江大丰实业股份有限公司（舞台设备等）<br>马歇尔戴声学有限公司 Marshall Day acoustics（声学顾问） |
| 项目摄影 | 存在建筑 |

**Shaanxi Opera House**

| Function | Cultural Building |
| --- | --- |
| Location | Xi'an, Shaanxi Province |
| Design / Completion | 2005 / 2017 |
| Floor Area | 52,324m$^2$ |
| Client | Xi'an Qujiang New District Social Service Center<br>Xi'an Qujiang Great Tang Everbright City Culture and Commerce (Group) Co., Ltd. |
| Design Team | Teng Luying, Zheng Ying, Wang Huan, Li Xiaojun, Wang Yuluo, Xu Jiangqi, Li Na, Huang Limiao, Qu Zilan, Li Jiang, Gao Kai, Gu Jindi, et al. |
| Collaborators | Northwest Architectural Design and Research Institute (Construction drawing design)<br>Zhejiang Dafeng Industrial Co., Ltd. (Stage equipment advisory, etc)<br>Marshall Day Acoustics Co., Ltd. (Acoustics advisory) |
| Photographer | Arch-Exist |

**宁波文化广场**

类型　文化商业建筑
位置　浙江省宁波市
时间　2008 / 2012
建筑面积　320 815m²
业主　宁波文化广场投资发展有限公司
设计参与　吴欣、董立军、滕露莹、史晨鸣、王瑜珞、颜莺、马庆禕、郑滢 等
合作设计　宁波市城建设计研究院有限公司（施工图设计）
项目摄影　吕恒中

**Ningbo Cultural Plaza**

Function　Cultural and Commercial Building
Location　Ningbo, Zhejiang Province
Design / Completion　2008 / 2012
Floor Area　320,815m²
Client　Ningbo Cultural Plaza Investment and Development Co., Ltd.
Design Team　Wu Xin, Dong Lijun, Teng Luying, Shi Chenming, Wang Yuluo, Yan Ying, Ma Qingyi, Zheng Ying, et al.
Collaborator　Ningbo Construction Design Institute Co., Ltd. (Construction drawing design).
Photographer　Lyu Hengzhong

**金华科技文化中心**

类型　文化建筑
位置　浙江省金华市
时间　2015 / 2019
建筑面积　81 480m²
业主　金华市多湖中央商务区建设投资有限公司
设计参与　马庆禕、肖俊瑰、颜莺、杨远、章墨、赵建霞、黄立妙、韩瑨玚、熊熙雯、杨涛（建筑设计团队）滕露莹、黄紫璇、邵帅、徐荣耀、李江、赵波（室内顾问设计团队）
合作设计　浙江省建筑设计研究院（施工图设计）
浙江大学建筑设计研究院有限公司-幕墙分院（幕墙设计）
浙江农林大学园林设计院有限公司（景观设计）
浙江永麒照明工程有限公司（泛光设计）
浙江亚厦装饰有限公司-苏州设计分院（室内设计）
尤尼可照明设计（北京）有限公司（室内照明设计）
南京曼式声学（声学设计）
项目摄影　存在建筑

**Jinhua Science & Culture Complex**

Function　Cultural Building
Location　Jinhua, Zhejiang Province
Design / Completion　2015 / 2019
Floor Area　81,480m²
Client　Jinhua Duohu Central Business District Construction Investment Co., Ltd.
Design Team　Ma Qingyi, Xiao Jungui, Yan Ying, Yang Yuan, Zhang Mo, Zhao Jianxia, Huang Limiao, Han Jinchang, Xiong Xiwen, Yang Tao(Architectural design team)
Teng Luying, Huang Zixuan, Shao Shuai, Xu Rongyao, Li Jiang, Zhao Bo（Interior advisory design team）
Collaborator　Zhejiang Institute of Architectural Design (Construction drawing design)
Architectural Design & Research Institute of Zhejiang University Co., Ltd. (Curtain wall design)
Zhejiang A&F University Landscape Design Institute Co., Ltd. (Landscape design)
Zhejiang YOKELIGHT Engineering Co., Ltd. (Floodlight design)
Zhejiang YASHA Decoration Co., Ltd. (Interior design)
UNICORE Lighting Design (Beijing) Co., Ltd. (Interior lighting design)
Nanjing Manshi Acoustics Co., Ltd. (Acoustic design)
Photographer　Arch-Exist

**西安碑林博物馆设计**

类型　文化建筑
位置　陕西省西安市
时间　2018 /
建筑面积　95 169m²
业主　西安城墙文化投资发展有限公司
设计参与　郑滢、肖俊瑰、瞿子岚、韩耀宗、彭智谦、陈溟澈 等

**Xi'an Beilin Museum**

Function　Cultural Building
Location　Xi'an City, Shaanxi Province
Design / Completion　2018 /
Floor Area　95,169m²
Client　Xi'an City Wall Culture Investment Development Co., Ltd
Design Team　Zheng Ying, Xiao Jungui, Qu Zilan, Han Yaozong, Peng Zhiqian, Chen Mingche, et al.

图书在版编目（C I P）数据

项秉仁建筑实践：1976–2018 / 滕露莹等编. -- 上海：同济大学出版社，2021.1
ISBN 978-7-5608-9507-9

Ⅰ. ①项… Ⅱ. ①滕… Ⅲ. ①建筑设计 - 作品集 - 中国 - 现代 Ⅳ. ①TU206

中国版本图书馆 CIP 数据核字(2020)第177013号

《项秉仁建筑实践1976–2018》编委会

滕露莹、马庆禕、蔡沪军、郑滢、颜莺、肖俊瑰、周希冉、曹佟、李芸、奚梓芙、赵阳、田园诗 等

前期策划 秦蕾 / 群岛工作室
英文翻译 杨碧琼、徐洲
图片摄影 存在建筑、曹呈祥、吕恒中、郭新新、张嗣烨、黄紫璇 等

ISBN 978-7-5608-9507-9
Publisher Hua Chunrong
Editor Chao Yan
Graphic Designer Fu Chao
Proofreader Xu Chunlian

Published in January 2021, by Tongji University Press,
1239, Siping Road, Shanghai, China, 200092.
www.tongjipress.com.cn


出版人 华春荣
责任编辑 晁艳
平面设计 付超
责任校对 徐春莲
版次 2021年1月第1版
印次 2021年1月第1次印刷
印刷 上海雅昌艺术印刷有限公司
开本 889mm × 1194mm 1/16
印张 19 插页3
字数 620 000
书号 ISBN 978-7-5608-9507-9
定价 328.00元
出版发行 同济大学出版社
地址 上海市杨浦区四平路1239号
邮政编码 200092
网址 http://www.tongjipress.com.cn
经销 全国各地新华书店
本书若有印装质量问题，请向本社发行部调换。


**Xiang Bingren Architectural Practice 1976–2018**

# 项秉仁 建筑实践1976 — 2018

滕露莹 马庆禕 曹佟 李芸 编
Teng Luying, Ma Qingyi, Cao Tong, Li Yun

luminocity.cn

光 明 城

LUMINOCITY

“光明城”是同济大学出版社城市、建筑、设计专业出版品牌，由群岛工作室负责策划及出版，致力以更新的出版理念、更敏锐的视角、更积极的态度，回应今天中国城市、建筑与设计领域的问题。